Perspectives of Elementary Mathematics

Gerhard P. Hochschild

Perspectives of Elementary Mathematics

Springer-Verlag
New York Berlin Heidelberg Tokyo

Gerhard P. Hochschild
Department of Mathematics
University of California
Berkeley, CA 94720
U.S.A.

AMS Classification: 00–01

Library of Congress Cataloging in Publication Data
Hochschild, Gerhard P. (Gerhard Paul), 1915–
Perspectives of elementary mathematics.
1. Mathematics—1961– . I. Title.
QA39.2.H617 1983 510 83-658

Typeset by Composition House Ltd., Salisbury, England.
Printed and bound by R. R. Donnelley & Sons, Harrisonburg, VA.
Printed in the United States of America.

9 8 7 6 5 4 3 2 1

ISBN 0-387-**90848**-X Springer-Verlag New York Berlin Heidelberg Tokyo
ISBN 3-540-**90848**-X Springer-Verlag Berlin Heidelberg New York Tokyo

Preface

Primarily, this book addresses beginning graduate students expecting to become mathematicians or mathematically oriented computer scientists. Accordingly, the presentation is conditioned in content as well as in form by the assumption that the reader has already made an internal commitment to mathematics and is seeking not only mathematical information but also active involvement with mathematical pursuits.

The general aim of what follows is to present basic mathematical concepts and techniques in familiar contexts in such a way as to illuminate the nature of mathematics as an art. Thus, the selection and organization of the material is based on considerations regarding the philosophical significance of various mathematical notions and results, their interdependence and their accessibility. In other words, this text concentrates on displaying coherent mathematical material meriting exceptionally careful and expansive contemplation. It should not be regarded as a reference for the most frequently used results or methods of elementary mathematics.

The exposition is meant to be reasonably self-contained and to embody a growth pattern of mathematical ideas (for which no *historical* validity is claimed, of course). In order to avoid burying the essentials under routine technicalities, a style has been adopted that relies on the reader's active involvement somewhat more than is customary in texts for beginners. The exercises proposed at the end of each chapter are frequently extensions of the chapter content, rather than mere illustrations. They are designed to be manageable in a straightforward fashion within the framework provided by the text.

The projects following the exercises are concerned with the creation of computer programs. However, for the purposes at hand, it is not necessary to do any formal programming. It suffices to think of a computer program in the informal sense, as a detailed recipe for accomplishing a computational task by a routine that, in principle, can be completely automated. The point to be made by the inclusion of such projects is that often the construction of a formal proof in a mathematical exposition is essentially the same enterprise as the creation of a computer program. At any rate, it is hoped that, in the contexts to which the projects refer, the programming enterprise will contribute to the clarification and appreciation of the underlying mathematics. Moreover, the reader with some programming expertise may find that the complete creation and running of the suggested programs is intellectually rewarding.

List of Chapters

CHAPTER I

Counting

1. The ritual of *counting* a collection of objects consists in attaching, at least mentally, a certain label to each object of the collection in turn. This presupposes that an ordered supply of labels, like (1, 2, 3, . . .), is available, and the labels are successively selected for use in the given order. The last label used is taken to be the measure of the *size* of the collection. The significance of this ritual resides in the following basic principles of set theory.

Intuitively, a set is a collection of objects, called *elements* of the set. In mathematics, no attempt is made to reduce this notion further, and "set" is adopted as a primitive, undefined notion. In particular, the elements of a set themselves are taken to be sets. What remains of the intuitive idea of collections consists in two notions of relationship among sets. If U and V are sets, it may be the case that U *is an element of* V. The customary indication of this relationship is $U \in V$. This is expressed also be saying that U *belongs to* V. The second relationship among sets is that of *containment*. One says that U *is contained in* V if every element of U is also an element of V. If this is the case, one calls U a *subset* of V, and one indicates this relationship by $U \subset V$. Motivated by the intuitive idea of a collection, one declares, as an axiom, that if U is contained in V and V is contained in U then U and V are identical: $U = V$.

The notion of set is quite literally fundamental in mathematics. Sets are the building blocks as well as the buildings of mathematics. If this is taken as a programmatic dictum, the first question one has to face is this. Without appealing to anything extraneous to mathematics, how can one even think of a single set? There is a peculiarly ingenious answer. If we have no objects before us, let us look at nothing! This provides us with a set. Namely, the set having no elements. One refers to this primitive set as the *empty* or *void* set, and one denotes it by $\varnothing$.

Once one has any set U, one may contemplate the set having U as its only element. This set is denoted by $\{U\}$. Now we have a new set, $\{\varnothing\}$, and this formation of sets can be iterated to yield a large supply of sets, each having just one element.

There are several other ways of constructing new sets from given ones. Given sets A and B, there is a set $A \cup B$, called the *union* of A and B; its elements are those of A and those of B. Another set coming from A and B is $A \cap B$, the *intersection* of A and B; its elements are those sets which belong to A as well as to B. Of course, the formations of union and intersection can be iterated. More generally, let S be any set. We wish to refer to the elements of the elements of S. In order to make this more intelligible, let us think of S as a *family* of sets and refer to its elements as *members*. Then the union of (the members of) S is the set whose elements are all those sets which belong to *some* member of S. This may be denoted by $\bigcup_{A \in S} A$, or simply by $\bigcup(S)$. Similarly, the intersection of S is the set whose elements are those sets which belong to *every* member of S. This is denoted by $\bigcap_{A \in S} A$, or simply by $\bigcap(S)$.

By forming unions, we can make sets having more than one element. For example, $\{\varnothing\} \cup \{\{\varnothing\}\}$ has the elements $\varnothing$ and $\{\varnothing\}$. This procedure is used in the set theoretical construction of the *natural numbers*, which usually serve as the labels in counting. One may define

$$0 = \varnothing, \qquad 1 = \{\varnothing\}, \qquad 2 = \{\varnothing\} \cup \{\{\varnothing\}\},$$

and generally the *successor* n' $(= n + 1)$ of a natural number n may be defined as $n \cup \{n\}$.

Next, if A is any set, we can form the set whose elements are the subsets of A. We denote this set by $P(A)$ and refer to it as the *set of subsets* of A. It is understood that the void set $\varnothing$ is a subset of every set, i.e., $\varnothing$ belongs to $P(A)$ for every set A. In particular, we have $P(\varnothing) = \{\varnothing\}$. For every element S of $P(A)$, there is an element T of $P(A)$ whose elements are precisely those elements of A which do not belong to S. One calls T the *complement* of S in A, and we denote it by $A \backslash S$.

The last set formation in our preliminary list is as follows. If A and B are sets, there is a set $A \times B$ whose elements are the *ordered pairs* (a, b) having as their first entry an element a of A and as their second entry an element b of B. This set is called the *Cartesian*, or *set theoretical product* of A and B. Here, the notion of ordered pair requires some theoretical underpinning, because "first" and "second" have only intuitive meanings. This difficulty is overcome by the definition $(a, b) = \{\{a\}\} \cup \{\{a\} \cup \{b\}\}$, which says that (a, b) is the set whose elements are $\{a\}$ and $\{a\} \cup \{b\}$. Note that all the "ordering" is to accomplish here is to ensure that the two entries play different roles, so that $(a, b) \neq (b, a)$ whenever $a \neq b$. It is easy to verify that the above definition of ordered pair satisfies this requirement, noting that if $a = b$ then (a, b) has $\{a\}$ as its only element, while otherwise (a, b) has exactly two elements, from which a and b are recoverable in *different* ways.

2. Mathematical life begins with the introduction of the concept of a *map* (short for mapping) or *function* from a set A to a set B. Informally, such a map is the assignment of an element of B to each element of A. The result of doing this may be viewed as a list consisting of ordered pairs (a, b) belonging to $A \times B$, where b is the element of B that has been assigned to the element a of A. This list contains exactly one ordered pair (a, b) for each element a of A. In other words, a map from A to B is a subset of $A \times B$ having the special property just mentioned.

If the map is named f, and if (a, b) is an ordered pair belonging to f, then the knowledge of the element a of A and the map f determines the second entry b of the ordered pair. For this reason, one simply writes $f(a)$ for b. Thus, our ordered pair is $(a, f(a))$. The element $f(a)$ of B is called the *value* of f at a. The set A is called the *domain* of the map f, and the subset of B whose elements are the values $f(a)$ is called the *image* of A in B by f, or simply the image of f. This image is frequently denoted by $f(A)$, although this is a misuse of the name f.

We say that f is *injective* if $f(x) \neq f(y)$ whenever $x \neq y$. We say that f is *surjective* if $f(A) = B$. If f is both injective and surjective then f is said to be *bijective*.

Most of mathematics consists in the study and manipulation of maps. The basic manipulative process is the *composition* of maps. Suppose that f is a map from a set A to a set B, and that g is a map from B to a set C. The *composite* of f with g is the map from A to C that is obtained as follows. Given an element a of A, first form the element $f(a)$ of B. Now g yields the element $g(f(a))$ of C. This composite is denoted $g \circ f$, and its formal definition is $(g \circ f)(a) = g(f(a))$ for every element a of A. Sometimes, one writes $g(f)$ for $g \circ f$.

Suppose we are also given a map h from C to a set D. Then we can form the composite $h \circ (g \circ f)$. On the other hand, we could first form $h \circ g$ and then $(h \circ g) \circ f$. Evidently, these two iterated composites are actually identical, i.e., the following *associative law* of composition is valid: $h \circ (g \circ f) = (h \circ g) \circ f$.

Although it is not our intention to enter into the technicalities of set theory, we must briefly discuss the classification of sets according to their size or *cardinality*. The most important size distinction is between *finite* sets and *infinite* sets. A tentative definition of "finite" consists in saying that a set is finite if it can be exhausted by throwing out its elements one at a time, i.e., if the ritual of counting it is effective. This can be made into an acceptable definition after developing the machinery of maps and a rigorous definition of natural numbers. On the other hand, there is an interesting way of avoiding such an appeal to more elaborate set theory.

Recall that, with a set S, we have associated the set $P(S)$ of its subsets. This set has an ordering coming from the relation of containment. Let us consider an arbitrary subset T of $P(S)$, i.e., a collection of subsets of S. Let us say that an element t of T is a *maximal* element of T if the only element u of T for which

$t \subset u$ is the element t itself. Then the following definition of "finite" can be shown to be in accord with the intuitive notion and to be equivalent to any of the other accepted definitions. We say that *a set S is finite if every non-empty subset of $P(S)$ has a maximal element*. By considering complements, we see that an equivalent definition is obtained by replacing "maximal" with "minimal," with the understanding that t is a *minimal* element of T if it is the only element u of T for which $u \subset t$.

Here is how the set N of all natural numbers fails to be finite. Recall that a natural number n is actually a subset of N, its elements being the predecessors of n. In this way, N appears as a subset of $P(N)$. The successor n' of a natural number n contains n as a subset distinct from n'. It is clear from this that, when viewed as a subset of $P(N)$, the set N has no maximal element.

Returning to our opening theme, we may now say that counting a finite set A means constructing an injective map c from A to the set N of natural numbers, with the property that $c(A)$ coincides, as a subset of N, with a natural number n. This property is the rigorous formulation of the "orderly-ness" of the counting process. In fact, via c, the set A is endowed with a total ordering, in which any two elements are comparable, corresponding to the ordering of $n = c(A)$ by containment. It is not an entirely trivial task to show that the natural number $c(A)$ depends only on A, and not on the particular choice of the map c. At any rate, if A and B are finite sets, then the corresponding natural numbers, i.e., the *cardinalities* of A and B, are the same if and only if there is a bijective map from A to B. Also, a set S is finite if and only if there is a bijective map from S to some natural number.

If A and B are finite sets of the same cardinality then a map from A to B is injective if and only if it is surjective. We remark that nothing like this holds for infinite sets: if A and B are infinite sets that are of the same size, in the sense that there is a bijective map from A to B, then there exist injective maps from A to B that are not surjective, as well as surjective maps from A to B that are not injective.

3. Let us review some of the elementary facts connected with counting finite sets. For every finite set A, we shall denote the cardinality of A by $\mathbf{N}(A)$. If A and B are finite sets, one has

$$\mathbf{N}(A) + \mathbf{N}(B) = \mathbf{N}(A \cup B) + \mathbf{N}(A \cap B)$$

and

$$\mathbf{N}(A \times B) = \mathbf{N}(A)\mathbf{N}(B)$$

Actually, the second of these equalities may be used for defining the multiplication of natural numbers, and the first equality may be used for defining the addition of natural numbers by taking A and B such that $A \cap B = \varnothing$.

Thus, if m and n are natural numbers, one may define

$$mn = \mathbf{N}(m \times n)$$

and

$$m + n = \mathbf{N}((\{0\} \times m) \cup (\{1\} \times n))$$

Let $M(A, B)$ denote the set whose elements are the maps from A to B. Such a map may be constructed by choosing, for each element of A in turn, an element of B as the value of the map at the given element of A. At each step, there are $\mathbf{N}(B)$ possible choices. Hence, we have

$$\mathbf{N}(M(A, B)) = \mathbf{N}(B)^{\mathbf{N}(A)}$$

Now let $I(A, B)$ denote the set whose elements are the *injective* maps from A to B. Proceeding as above, we note that, after we have chosen values for k elements of A, there are $\mathbf{N}(B) - k$ possible choices for the next value, as long as k does not exceed $\mathbf{N}(B)$. This yields the result

$$\mathbf{N}(I(A, B)) = \mathbf{N}(B)(\mathbf{N}(B) - 1)\cdots(\mathbf{N}(B) - \mathbf{N}(A) + 1)$$

Note that the product on the right equals 0 whenever $\mathbf{N}(B)$ is smaller than $\mathbf{N}(A)$, in which case there are no injective maps from A to B. An empty product is to be interpreted as 1.

From this last result, we can obtain the number of subsets of B having exactly k elements. Let us write $\mathbf{N}(B) = n$, and let A be any set having exactly k elements. Let S be a subset of B such that $\mathbf{N}(S) = k$. By our last result, the number of injective maps from A to S is $k(k - 1)\cdots 1 = k!$. In other words, the number of injective maps from A to B with image S is equal to $k!$. Now let n_k denote the number of subsets of B having exactly k elements. Then it is clear from what we have just seen that the number of injective maps from A to B is equal to $k!\,n_k$. On the other hand, we know that this number is equal to $n(n - 1)\cdots(n - k + 1)$. Thus, the number of subsets of B having exactly k elements is equal to $n(n - 1)\cdots(n - k + 1)/k!$. Of course, this is the *binomial coefficient*

$$\binom{n}{k} = n!/(k!\,(n - k)!)$$

Finally, let $S(A, B)$ denote the set whose elements are the *surjective* maps from A to B. Such a map, f say, is determined when it is known, for each element b of B, which elements of A are sent to b by f. Accordingly, let us denote the set of those elements a of A for which $f(a) = b$ by $f^{-1}(b)$. Now let $b_1, \ldots, b_n$ be the elements of B. Then the sets $f^{-1}(b_1), \ldots, f^{-1}(b_n)$ are pairwise *disjoint*, i.e., if $i \neq j$ then $f^{-1}(b_i) \cap f^{-1}(b_j) = \varnothing$. The union $f^{-1}(b_1) \cup \cdots \cup f^{-1}(b_n)$ coincides with A. Put $k_i = \mathbf{N}(f^{-1}(b_i))$ and $m = \mathbf{N}(A)$. Then each k_i is positive, and $k_1 + \cdots + k_n = m$.

Let us momentarily fix such a set of k_i's and determine the number of maps f such that $\mathbf{N}(f^{-1}(b_i)) = k_i$ for each i. There are $\binom{m}{k_1}$ choices for $f^{-1}(b_1)$. Once $f^{-1}(b_1)$ has been chosen, the next set $f^{-1}(b_2)$ is a subset of $A \setminus f^{-1}(b_1)$. This last set has $m - k_1$ elements, so that the number of choices for $f^{-1}(b_2)$ is $\binom{m-k_1}{k_2}$. Continuing this argument, we see that the number of such maps f is $\binom{m}{k_1} \cdots \binom{m - k_1 - \cdots - k_{n-1}}{k_n}$. Writing these binomial coefficients in terms of factorials and then making the evident cancellations yields the result $\frac{m!}{k_1! \cdots k_n!}$. Hence, we have

$$\mathbf{N}(S(A, B)) = \sum \frac{m!}{k_1! \cdots k_n!}$$

where the summation $\sum$ is understood to go over all ordered n-tuples $(k_1, \ldots, k_n)$ of positive natural numbers such that $k_1 + \cdots + k_n = m$. If the number n of elements of B is greater than the number m of elements of A there are no such n-tuples, and the sum is to be interpreted as 0.

4. Here is an interesting problem involving counts of sets in a subtle way. Let B and S be finite sets, and suppose there is given a map η from B to $P(S)$. We ask if there is an *injective* map f from B to S such that $f(b)$ belongs to the subset $\eta(b)$ of S for every element b of B. Clearly, for such an f to exist, η must satisfy the following condition. For every subset C of B, the union of the family of sets $\eta(c)$ with c in C must have at least as many elements as C. We shall show that this condition is actually sufficient to ensure the existence of a map f satisfying the above requirements.

Let $\eta'(C)$ denote the union of the family of sets $\eta(c)$, with c ranging over C. Then our condition may be written $\mathbf{N}(\eta'(C)) \geqq \mathbf{N}(C)$ for every subset C of B. In the case where $B = \varnothing$, the result holds trivially, with f the empty map. Clearly, the result holds also in the case where $\mathbf{N}(B) = 1$. Therefore, we suppose that $\mathbf{N}(B) > 1$.

Let us say that a subset Q of B is *good* if the result we wish to establish for the pair (B, S) actually holds for the pair (Q, S). If B is good we are done. Otherwise the subset of $P(B)$ whose elements are the subsets of B that are not good is non-empty. Since B is finite, this family has a minimal element, M say. Now every *proper* subset of M, i.e., every subset of M other than M itself, is good. Therefore, it suffices to prove the result in the case where every proper subset of B is good, which we shall now assume to be the case.

First, we deal with the situation where there is a proper subset C of B such that $C \neq \varnothing$ and $\mathbf{N}(\eta'(C)) = \mathbf{N}(C)$. By our assumption, there is an injective map g from C to S such that $g(c)$ belongs to $\eta(c)$ for every element c of C. Using η, we construct a map τ from $B\backslash C$ to $P(S)$ by putting $\tau(d) = \eta(d) \cap (S\backslash\eta'(C))$ for every element d of $B\backslash C$. We show that τ satisfies the condition of our problem.

Let E be any subset of $B\backslash C$. Since $E \cap C = \varnothing$, we have

$$\mathbf{N}(E \cup C) = \mathbf{N}(E) + \mathbf{N}(C).$$

By the assumption concerning η, we have $\mathbf{N}(\eta'(E \cup C)) \geqq \mathbf{N}(E \cup C)$. The definition of τ shows that $\tau'(E) \cup \eta'(C) = \eta'(E \cup C)$ and $\tau'(E) \cap \eta'(C) = \varnothing$. Hence, $\mathbf{N}(\eta'(E \cup C)) = \mathbf{N}(\tau'(E)) + \mathbf{N}(\eta'(C))$. It follows that we have $\mathbf{N}(\tau'(E)) + \mathbf{N}(\eta'(C)) \geqq \mathbf{N}(E) + \mathbf{N}(C)$. Since $\mathbf{N}(\eta'(C)) = \mathbf{N}(C)$, this gives the required result $\mathbf{N}(\tau'(E)) \geqq \mathbf{N}(E)$.

Since $B\backslash C$ is good, we may now conclude that there is an injective map h from $B\backslash C$ to S such that $h(d)$ belongs to $\tau(d)$ for every element d of $B\backslash C$. This implies that $h(d)$ belongs to $\eta(d)$, but *not* to $\eta'(C)$. Now we define a map f from B to S by putting $f(x) = g(x)$ if x belongs to C, and $f(x) = h(x)$ if x belongs to $B\backslash C$. Since g has all its values in $\eta'(C)$, while no value of h belongs to $\eta'(C)$, the fact that g and h are injective implies that f is injective. Evidently, $f(x)$ belongs to $\eta(x)$ for every element x of B. Thus, f satisfies all the requirements.

It remains to deal with the case where, for every non-empty proper subset C of B, one has $\mathbf{N}(\eta'(C)) > \mathbf{N}(C)$. In this case, we select any element b from B and an element s from $\eta(b)$. Next, we define a map ρ from $B\backslash\{b\}$ to $P(S)$ by $\rho(x) = \eta(x) \cap (S\backslash\{s\})$. Here, it is evident that ρ satisfies the condition of our problem. Since $B\backslash\{b\}$ is good, we may therefore conclude that there is an injective map g from $B\backslash\{b\}$ such that $g(x)$ belongs to $\rho(x)$ for every element x of B other than b. Finally, we define a map f from B to S by putting $f(b) = s$ and $f(x) = g(x)$ for every element x of B other than b. Evidently, f satisfies the requirements for a solution of our problem, so that the proof is now complete.

For a reason the reader may discover, the result we have just proved is called the *marriage lemma*. An application of it in a different direction is the following.

Suppose that a finite set A is presented as a union of a family of subsets in two ways:

$$B_1 \cup \cdots \cup B_n = A = C_1 \cup \cdots \cup C_n$$

where each B_i and each C_i has exactly k elements, and $B_i \cap B_j = \varnothing = C_i \cap C_j$ whenever $i \neq j$. We shall prove that there are elements $a_1, \ldots, a_n$ in A such that each B_i and each C_i contains exactly one of the a_j's.

Let T_i be the set whose elements are those C_j's for which $C_j \cap B_i \neq \varnothing$. The union of any family of p of the B_i's has exactly pk elements. Therefore, such a union cannot be contained in the union of a family of fewer than p of the C_j's. This implies that the union of the family of the corresponding p sets

T_i consists of at least p sets C_j. This means that the condition of the marriage lemma is satisfied by the map η, where $\eta(B_i) = T_i$. It follows from the lemma that there is an injective map f from the set of the B_i's to the set of the C_i's such that $f(B_i) \cap B_i \neq \varnothing$ for each i. Let a_i be any element of $f(B_i) \cap B_i$. Then $a_1, \ldots, a_n$ are as required.

One can improve this result by iteration, finally obtaining

$$A = A_1 \cup \cdots \cup A_k$$

with $A_r \cap A_s = \varnothing$ whenever $r \neq s$, and $\mathbf{N}(A_r \cap B_i) = 1 = \mathbf{N}(A_r \cap C_i)$ for all r and i.

5. We have seen in Section 2 that the set N of natural numbers is totally ordered by the containment (or inclusion) relation. Generally, an *ordering* of a set A is a relation among elements of A, usually indicated by the sign $\leqq$, having the following properties. For every element x of A, one has $x \leqq x$. If x and y are elements of A such that $x \leqq y$ and $y \leqq x$ than $x = y$. Finally, if x, y and z are elements of A such that $x \leqq y$ and $y \leqq z$ then $x \leqq z$. Such an ordering is called a *total* ordering if, for all elements x and y of A, either $x \leqq y$ or $y \leqq x$.

If f is an injective map from A to an ordered set B then f establishes an order in A by the agreement that $x \leqq y$ means $f(x) \leqq f(y)$. We have already met an example of an ordering obtained in this way, where A is a finite set, $B = N$ and f is a counting map. In general, *having an ordering on a set A is equivalent to having an injective map f from A to $P(A)$*. Indeed, if A is endowed with an ordering, we can define f by making each value $f(a)$ the subset of A consisting of all elements b such that $b \leqq a$. Evidently, f is injective, and the given ordering of A coincides with the ordering established by f from the natural ordering of $P(A)$ by the containment relation. Conversely, from an injective map f from A to $P(A)$, we can define an ordering of A by the agreement that $a \leqq b$ means $f(a) \subset f(b)$. Note that the definition of ordered pair is based on this idea!

In order to proceed, we need some additional terminology concerning maps. If A is a set, one has the *identity map* from A to A. Denoting this by i_A, we have $i_A(x) = x$ for every element x of A. Suppose that f is a bijective map from a set A to a set B. Then the subset of $B \times A$ consisting of the ordered pairs (b, a) such that (a, b) belongs to f (i.e., such that $f(a) = b$) is a map from B to A. We denote this map by f^{-1} and refer to it as the *inverse* of f. It is clear from the definition that $f^{-1} \circ f = i_A$ and $f \circ f^{-1} = i_B$. A bijective map from A to A is called a *permutation* of A.

Let A be an arbitrary set that is endowed with a total ordering $\leqq$. Then every permutation ρ of A yields a total ordering $\leqq^\rho$, where $x \leqq^\rho y$ is defined to mean $\rho(x) \leqq \rho(y)$. In the case where A is a finite set, it is clear that these orderings $\leqq^\rho$, with ρ ranging over the set of all permutations of A, are all the total orderings of A. On the other hand, this is not true when A is an infinite set. For example, the set N of natural numbers, with its natural ordering $\leqq$, has 0 as a minimal element, i.e., $0 \leqq n$ for every element n of N. Consequently, if ρ is any permutation of N, then $\rho^{-1}(0)$ is a minimal element for the ordering $\leqq^{\rho}$. However, if we define the relation $\leqq^*$ by agreeing that $x \leqq^* y$ means $y \leqq x$, then $\leqq^*$ is a total ordering for which there is no minimal element.

A large part of mathematical procedure consists in establishing suitable orderings of sets, and the task of defining an ordering that is effective for the purpose at hand is often quite difficult, even when the set is finite. In fact, we are facing an ordering task right here. Namely, that of defining a manageable total ordering of the set of all permutations of a finite set.

Let us take our set to be the set of the natural numbers $1, 2, \ldots, n$, where $n > 1$, endowed with its natural order. For each natural number k with $2 \leqq k \leqq n$, let ρ_k denote the permutation of our set that acts by permuting the first k elements cyclically and leaves the remaining elements fixed. Thus, $\rho_k(i) = i$ if $i > k$, $\rho_k(k) = 1$ and $\rho_k(i) = i + 1$ if $i < k$. Let us indicate composition of permutations simply by juxtaposition, and let us use the exponent notation for composites of equal component permutations, agreeing that if ρ is a permutation of our set then ρ^0 stands for the identity map. Since $(\rho_k)^k$ is the identity map, while no smaller positive exponent here yields the identity map, the different powers of ρ_k are the $(\rho_k)^e$'s with $0 \leqq e < k$.

Consider the family of composites $(\rho_n)^{e_n} \cdots (\rho_2)^{e_2}$, where $0 \leqq e_k < k$ for each k. We claim that these are pairwise distinct. In order to see this, suppose that the same permutation is obtained with exponents f_k. Without loss of generality, we assume that $e_n \geqq f_n$. Composing with $(\rho_n)^{-1}$ on the left f_n times, we obtain the equality of the two composites with exponents $e_n - f_n$, $e_{n-1}, \ldots, e_2$ and $0, f_{n-1}, \ldots, f_2$. If $e_n > f_n$ then the first of these sends the element n of our set to $e_n - f_n$, while the second composite leaves n fixed. Consequently, we must have $e_n = f_n$. Now we are in the situation where $n - 1$ has taken the place of n. Iteration of this argument shows that $e_k = f_k$ for each k.

Thus, the number of distinct composites as written above is precisely $n!$, so that these composites constitute the set of *all* permutations of $(1, \ldots, n)$. Indexing these permutations by the corresponding $(n - 1)$-tuples $(e_2, \ldots, e_n)$ of exponents, we have a total ordering of the set of permutations, which is of the familiar *lexicographic* type.

Exercises

1. For arbitrary sets A, B, S, exhibit a bijective map from $M(A, M(B, S))$ to $M(A \times B, S)$, as well as its inverse.

2. Exhibit a bijective map from $M(S, A \times B)$ to $M(S, A) \times M(S, B)$, as well as its inverse.

3. Let f be a map from a non-empty set A to a non-empty set B. Show that f is injective if and only if there is a map g from B to A such that $g \circ f = i_A$. Try to get a similar characterization of surjective maps.

4. Let S be an arbitrary set, J a set with exactly two elements. Show that there is a bijective map from $P(S)$ to $M(S, J)$. Taking S finite, with exactly n elements, deduce that $\sum_{k=0}^{n} \binom{n}{k} = \mathbf{N}(P(S)) = 2^n$.

5. Let A, B, S be finite sets, with $A \cap B = \varnothing$. By counting $M(S, A \cup B)$ in a suitable manner, prove that $(\alpha + \beta)^\sigma = \sum_{\tau=0}^{\sigma} \binom{\sigma}{\tau} \alpha^\tau \beta^{\sigma-\tau}$, where $\alpha = \mathbf{N}(A)$, $\beta = \mathbf{N}(B)$, $\sigma = \mathbf{N}(S)$.

6. Let S be a set, and let f be an injective map from S to S. Suppose that, for every element x of S, the set of iterated images $f^n(x)$, where n ranges over N, is finite. Prove that f is bijective.

7. For any positive natural number n, let N^n denote the set of ordered n-tuples $(t_1, \dots, t_n)$ of natural numbers. Define a *path* in N^n as a finite sequence $\sigma_0, \dots, \sigma_q$ of such n-tuples such that each σ_{i+1} results from σ_i by adding 1 to exactly one of the entries. Let $a_1 \dots, a_n$ be natural numbers. Show that the number of paths starting at $(0, \dots, 0)$ and ending at $(a_1, \dots, a_n)$ is equal to $(a_1 + \cdots + a_n)!/(a_1! \cdots a_n!)$. [Hint: look at the determination of $\mathbf{N}(S(A, B))$ in Section 3.]

8. Let A and B be finite sets, with $\mathbf{N}(A) = m \geqq n = \mathbf{N}(B)$. For each index q with $0 \leqq q < n$, let $T(q)$ denote the number of maps f from A to B such that $\mathbf{N}(f(A)) = n - q$. Show that for each index p with $0 \leqq p < n$, one has

$$\sum_{q=p}^{n-1} \binom{q}{p} T(q) = \binom{n}{p} (n-p)^m$$

and hence that $T(0)$, i.e., the number of surjective maps from A to B, is equal to

$$\sum_{p=0}^{n-1} (-1)^p \binom{n}{p} (n-p)^m$$

[Hint: for the first part, consider the set of maps f from A to B for which $\mathbf{N}(f(A)) \leqq n - p$; for the second part, substitute for the summands the sums in the first formula, and then invert the order of summation.]

9. In the context and notation of the end of Section 5, show that $\rho_n \cdots \rho_2$ is the reversal, sending $(1, \dots, n)$ to $(n, \dots, 1)$. Denoting this reversal by τ, show that

$$\tau(\rho_n)^{e_n} \cdots (\rho_3)^{e_3} = (\rho_n)^{n+1-e_n} \cdots (\rho_3)^{4-e_3} \rho_2$$

Conclude from this that the set of $n!/2$ permutations $(\rho_n)^{e_n} \cdots (\rho_3)^{e_3}$ is a complete system of representatives for the set of reversal pairs $(\rho, \tau\rho)$. [Hint: first, show that $\tau\rho_n = (\rho_n)^{-1}\tau$.]

Projects

1. Suppose that A is a totally ordered finite set of labels. Write the subsets of cardinality p of A in the form $\{x_1, \ldots, x_p\}$, where the x_i's stand for the elements of the subset, and the indexing is in accord with the ordering of A. Think of arranging these elements of $P(A)$ in the corresponding lexicographic order, so that $\{x_1, \ldots, x_p\}$ precedes $\{y_1, \ldots, y_q\}$ if either $p < q$, or else $p = q$ and the lowest indexed x_i differing from y_i precedes y_i in the ordering of A. Devise a computer program that takes an element of $P(A)$ as input, in the form described above, and produces the immediate lexicographic successor as output.

2. Refer to Project 1 and to the proof of the marriage lemma in order to construct a recursive computer program for solving the marriage problem (B, S, η). The program might begin with using the program of Project 1 to search for a non-empty proper subset C of B for which $\mathbf{N}(\eta'(C)) = \mathbf{N}(C)$. If such a C is found, the program calls on itself with the input (C, S, η_C) (where η_C is the *restriction* of η to C, in the evident sense), for which the output is a map g from C to S, as in the proof of the marriage lemma. Next, the program calls on itself with the input $(B\backslash C, S\backslash\eta'(C), \tau)$ (where the notation is that of the proof of the marriage lemma), for which the output is a map h from $B\backslash C$ to $S\backslash\eta'(C)$, as used in the cited proof. Now the program simply combines g and h to produce the required map f. If no C as above exists and if B is non-empty, the program calls on itself with input $(B\backslash\{b\}, S\backslash\{s\}, \rho)$ (cf. Section 4), etc.

CHAPTER II

Integers

1. When the system of natural numbers is enlarged so as to yield richer number systems with a wider range of applicability, the first step leads to the system of *integers*. The gain in computational control obtained from this resides in the fact that, in the system of integers, the operation of adding a fixed number is invertible. In mundane terms, the new facility is the accountability of debts.

Formally, the system of natural numbers is a *monoid*, i.e., a system consisting of a set M and a map σ from $M \times M$ to M that is *associative*, in the sense that it satisfies the identity

$$\sigma(x, \sigma(y, z)) = \sigma(\sigma(x, y), z)$$

and for which there is a *neutral element*, e say, such that

$$\sigma(x, e) = x = \sigma(e, x)$$

for every element x of M. In the case where M is the system of natural numbers, we have $\sigma(x, y) = x + y$ and $e = 0$.

The system of integers is a *group*, by which is meant a monoid as above with the extra feature that, for every element x of M, there is an *inverse* x', characterized by the property $\sigma(x, x') = e = \sigma(x', x)$. In the general case, x' is usually denoted by x^{-1}. In the case of the system Z of integers, $x' = -x$.

The map σ is called the *composition* of the group or monoid, and usually one abbreviates $\sigma(x, y)$ by xy, provided there is no danger of confusion. In the case of the group Z of integers, there would result a disastrous confusion, since the group composition is addition, not multiplication.

If H is a subset of a group G such that the neutral element belongs to H and H is stable under the composition and inversion maps of G (i.e., if x and y

belong to H, so do xy and x^{-1}) then H evidently inherits a group structure from G by restriction. Such a group H is called a *subgroup* of G. In studying a given group G, the first problem one faces is that of determining its subgroups. In particular, the first problem of integer arithmetic is the determination of the subgroups of Z. This is easily solved, as follows.

Let a be any integer. Evidently, the set of all finite sums whose only summands are a or $-a$ is a subgroup of Z. Its elements are all the multiples xa, where x ranges over Z. Accordingly, we denote this subgroup by Za. Clearly, Za coincides with $Z(-a)$. Therefore, every such subgroup is a Za with $a \geqq 0$. Also, if a is positive, then it is the *smallest* positive element of Za. Consequently, if a and b are non-negative integers such that $Za = Zb$, then $a = b$. We shall show that every subgroup of Z is such a Za.

Let H be any subgroup of Z. If H does not consist of 0 alone, then H contains some positive integers. Assuming $H \neq (0)$, let h be the smallest positive element of H. Evidently, $Zh \subset H$. Conversely, let x be any positive element of H. By adding $-h$ repeatedly, starting with x, we eventually obtain either 0 or a positive integer smaller than h. Since all the integers generated by our additions belong to H, the second possibility is ruled out. Our conclusion is that x belongs to Zh. Finally, if y is a negative element of H, then $-y$ is a positive element of H, so that $-y$ belongs to Zh. But this implies that y belongs to Zh. Thus, we have $H = Zh$.

2. The above result concerning the subgroups of Z is actually the essence of the divisibility theory for integers. If a and b are integers the statement that *b divides a*, or that a is a *multiple* of b, is equivalent to the statement $Za \subset Zb$. This evident reformulation greatly enhances control of divisibility, as follows.

Let a and b be integers, and consider the set $Za + Zb$ of all sums $u + v$, with u in Za and v in Zb. Evidently, this is a subgroup of Z. Therefore, we have $Za + Zb = Zd(a, b)$, where $d(a, b)$ is a non-negative integer determined by a and b. It is easy to see that $d(a, b)$ is nothing but the *greatest common divisor* of a and b. Actually, $d(a, b)$ is the *smallest* positive integer of the form $xa + yb$ with x and y in Z, provided that at least one of a or b is not 0.

The algorithm for finding the greatest common divisor of two integers a and b (which we assume to be positive without losing anything essential) is the *Euclidean algorithm* of division. In this connection, it should be noted that the relevant operation of "division" is simply iterated addition of the negative of the "denominator." This becomes quite clear when one puts the greatest common divisor routine into the following recursive form.

Assuming that $a \geqq b > 0$, put $a_0 = a$ and $b_0 = b$. Given $a_n > b_n > 0$, if $-b_n + a_n \geqq b_n$ put $a_{n+1} = -b_n + a_n$ and $b_{n+1} = b_n$, otherwise put $a_{n+1} = b_n$ and $b_{n+1} = -b_n + a_n$. Stop the process when $a_n = b_n$, which must eventually happen, as is easy to see. Evidently, if $a_i, b_i, a_{i+1}, b_{i+1}$ are defined, we have $Za_i + Zb_i = Za_{i+1} + Zb_{i+1}$. Therefore, for the final index n, where $a_n = b_n = d$ say, we have $Zd = Za + Zb$, showing that $d = d(a, b)$.

It is clear that the intersection of any family of subgroups of some fixed group G is again a subgroup of G. In particular, with a and b as above, the intersection $(Za) \cap (Zb)$ is a subgroup of Z. It is therefore a $Zm(a, b)$, where $m(a, b)$ is a certain positive integer determined by a and b. In fact, $m(a, b)$ is the *least common multiple* of a and b.

3. In pursuing the above topic further, we shall use some general concepts from group theory. The first of these is that of a *group homomorphism.* What is meant by this is a map, η say, from a group G to a group H that respects the group compositions, in the sense that $\eta(xy) = \eta(x)\eta(y)$ for all elements x and y of G. The set of all elements x of G such that $\eta(x)$ is the neutral element of H is evidently a subgroup of G. It is called the *kernel* of η. The significance of this kernel, K say, resides in the fact that η is injective if and only if K is the *trivial* subgroup of G, whose only element is the neutral element. On the other hand, η also determines a subgroup of H; namely, $\eta(G)$. If η is bijective, i.e., if K is trivial and $\eta(G) = H$, then η is called a group *isomorphism.* Its inverse η^{-1} is easily seen to be a group isomorphism also.

Let S be any subgroup of G. For every element x of G, let xS stand for the subset of G whose elements are the composites xs with s in S. These subsets are called the *cosets* of S in G. If x and y are elements of G then either $xS = yS$ or else $(xS) \cap (yS) = \varnothing$. In other words, the cosets of S in G are pairwise disjoint. Evidently, G is the union of the family of these cosets. We shall denote the set of cosets of S in G by G/S. Note that, in the case where G is finite, what we have just seen implies that $\mathbf{N}(G) = \mathbf{N}(G/S)\mathbf{N}(S)$. The same remarks apply to the cosets Sx of the other kind. In the cases of principal interest to us at this stage, the composition of G satisfies the commutative law $xy = yx$, which is signalized by saying that G is a *commutative* group. In these cases, the two kinds of cosets coincide, of course.

In general, the subgroup S of G is called a *normal* subgroup if $xS = Sx$ for every element x of G. If S is a normal subgroup, and U and V are cosets of S in G, then the set UV of all composites uv with u in U and v in V is still a coset of S in G. Thus, the composition map of G yields a composition map for G/S. It is seen immediately that this makes G/S into a group. On calls this group the *factor group* of G with respect to S.

In particular, the kernel K of a group homomorphism η from G to H is always a normal subgroup of G, so that one can form the factor group G/K. The elements xK of G/K are precisely the *inverse images* $\eta^{-1}(\eta(x))$ of the elements $\eta(x)$ of $\eta(G)$, i.e., we have $\eta(x) = \eta(y)$ if and only if y belongs to xK. Consequently, η yields an injective map η' from G/K to H, where $\eta'(xK) = \eta(x)$. Evidently, η' is a group homomorphism. The map, π say, that is defined by $\pi(x) = xK$ for every x in G is clearly a surjective group homomorphism from G to G/K, which we shall call the *canonical* homomorphism. Of course, this applies to every normal subgroup K of G. Here, where K is the kernel of η, we have $\eta' \circ \pi = \eta$. More generally, *if μ is any group homomorphism from*

G to a group Q such that the kernel of μ contains the normal subgroup K of G, then there is one and only one group homomorphism μ' from G/K to Q such that $\mu' \circ \pi = \mu$.

Finally, if A and B are groups, one makes the Cartesian product set $A \times B$ into a group by defining the composition so that $(u, v)(x, y) = (ux, vy)$. This group is called the *direct product* of the groups A and B.

4. Let A be a commutative group, and indicate the group composition of A by $+$. Let U and V be subgroups of A. These determine two more subgroups of A, one of which is the intersection $U \cap V$, while the other is the group $U + V$ whose elements are the composites $u + v$ with u in U and v in V. We form the direct product $A \times A$ and consider the group homomorphisms δ from A to $A \times A$ and τ from $A \times A$ to A, where

$$\delta(x) = (x, x) \quad \text{and} \quad \tau(x, y) = -x + y$$

One verifies directly that δ is injective, τ is surjective, and the kernel of τ coincides with the image of δ. This very transparent system of group homomorphisms gives rise to a more subtle one, with the same composition pattern, involving U and V. Thus, δ yields a group homomorphism ρ from $A/(U \cap V)$ to $(A/U) \times (A/V)$, while τ yields a group homomorphism σ from $(A/U) \times (A/V)$ to $A/(U + V)$. The definitions of ρ and σ are based on the above homomorphism principle of factor groups. We exhibit ρ and σ in the following formulae, where we use the appropriate coset notation.

$$\rho(x + U \cap V) = (x + U, x + V)$$
$$\sigma(x + U, y + V) = -x + y + U + V$$

It is not difficult to verify that ρ is injective, σ is surjective, and the kernel of σ coincides with the image of ρ.

Now let us consider the case where A is the group Z of integers, and U and V are the subgroups Za and Zb corresponding to positive integers a and b. Then, in the notation of Section 2, we have $U \cap V = Zm(a, b)$ and $U + V = Zd(a, b)$. For every positive integer c, the number of elements of Z/Zc is equal to c. From the fact that ρ is injective, σ surjective and the kernel of σ coincides with the image of ρ, it follows that the cardinality of $(A/U) \times (A/V)$ is equal to the product of the cardinalities of $A/(U \cap V)$ and $A/(U + V)$. This says that $ab = m(a, b)d(a, b)$.

5. Several times in the above discussion, the second composition map of Z, the multiplication, has already come into play, although we concentrated as much as possible on the group structure of Z, the addition. The presence of two composition maps that interact in a somewhat subtle manner can be traced back to the context of group theory, where the interaction between "addition" and "multiplication" arises quite naturally.

Let A be a commutative group, whose composition map we call addition and indicate by $+$. A group homomorphism from A to A is called an *endomorphism* of A. Accordingly, we write End(A) for the set of all group homomorphisms from A to A. The addition of A yields a composition map for End(A), which again we call addition and indicate by $+$. If f and g are endomorphisms of A, their sum $f + g$ is defined as a map from A to A by the formula

$$(f + g)(x) = f(x) + g(x)$$

Using that the addition in A is commutative and associative, one verifies directly that $f + g$ is actually an endomorphism of A. Next, one verifies easily that this addition of endomorphisms makes End(A) into a commutative group.

On the other hand, the ordinary composition of maps makes End(A) into a monoid, whose neutral element is the identity map i_A. Except for the fact that the monoid composition of End(A) need not be commutative, the formal features of the full structure of End(A) are exactly those governing addition and multiplication of integers. Thus, as regards the interaction between addition and multiplication (here, the ordinary composition of maps), we have the two *distributive laws*

$$f \circ (g + h) = f \circ g + f \circ h \quad \text{and} \quad (f + g) \circ h = f \circ h + g \circ h$$

In general, a system consisting of a commutative group (written additively) that is endowed with a second composition map for which it is a monoid (written multiplicatively) is called a *ring*, provided that the two distributive laws displayed above are valid. The important maps from a ring R to a ring S are the *ring homomorphisms*, i.e., the group homomorphisms, h say, that respect also the multiplication, in the sense that $h(xy) = h(x)h(y)$ for all elements x and y of R, and that h sends the neutral element 1_R (or just 1) for the multiplication of R to the neutral element for the multiplication of S.

Observe that the distributive laws for a ring R mean that the multiplication by a fixed element, from either side, is a group homomorphism from R to R. Moreover, if μ is the map from R to End(R) (the ring of *group*-endomorphisms of R) that sends each element x to the multiplication by x from the left (i.e., $\mu(x)(y) = xy$) then μ is a ring homomorphism. It is injective, as is seen by noting that $\mu(x)(1) = x$.

Now let us return to the group Z of integers. When equipped with the multiplication, Z is a ring. Since the multiplication by a positive integer is just iterated addition of the multiplicand, the ring structure of Z is already determined by its group structure. However, this feature can be exhibited with more precision in the terms introduced just above. In the case where $R = Z$, the above ring homomorphism μ from Z to End(Z) is actually bijective, so that *the ring Z appears as a copy of the natural ring* End(Z).

The fact that μ is bijective is seen as follows. We have already seen in the general case that μ is injective. Now, if f is any element of End(Z), it is easy to show inductively that f is the multiplication by the element $f(1)$, i.e., that $f = \mu(f(1))$.

Let R be an arbitrary ring, and let T be a subgroup of R. We may form the factor group R/T, and there arises the question of the conditions under which the factor group inherits a multiplication from R such that R/T becomes a ring and the canonical group homomorphism from R to R/T becomes a ring homomorphism. It is seen almost immediately that this is the case provided only that the left and right multiplications by arbitrary elements of R send T into T, i.e., that $xT \subset T$ and $Tx \subset T$ for every element x of R. In this case, one calls T an *ideal* of R, and one refers to R/T as the *factor ring* of R with respect to T. Note also that the kernel of every ring homomorphism is an ideal. The fact that the multiplication of Z comes from the addition is reflected in the property that every subgroup of Z is actually an ideal, so that every factor *group* of Z is actually a factor *ring*.

6. Let F be a commutative ring, i.e., a ring whose multiplication is commutative. We write 1 for the neutral element of the multiplication of F, and we assume that $1 \neq 0$ or, equivalently, that F has more than one element. If the multiplication of F makes the set of elements other than 0 into a group, one calls F a *field*. This last requirement is simply that, for every non-zero element x of F, there is a *reciprocal* x^{-1} such that $x^{-1}x = 1$.

Suppose that R is a commutative ring, and let T be a proper ideal of R, i.e., an ideal other than R. Consider the factor ring R/T. If ρ denotes the canonical ring homomorphism from R to R/T then the ideals of R/T are precisely the images $\rho(J)$, where J ranges over the set of those ideals of R which contain T. It is easy to see that a commutative ring other than (0) is a field if and only if (0) is its only proper ideal. Consequently, R/T is a field if and only if the only ideals containing T are T and R, i.e., if and only if T is maximal in the set of all proper ideals of R. In this case, we call T simply a *maximal ideal*.

Let t be an integer greater than 1, so that Zt is a proper ideal of Z. From the above, we know that Z/Zt is a field if and only if Zt is maximal in the set of proper ideals of Z, which set is simply the set of proper subgroups of Z. Evidently, Zt is a maximal proper subgroup of Z if and only if the only positive divisors of t are t and 1, i.e., if and only if t is a prime number. Thus, *the prime numbers are precisely those positive integers for which Z/Zt is a field.*

Let p be a prime number, and let π denote the canonical ring homomorphism from Z to Z/Zp. Then an integer u is divisible by p if and only if $\pi(u) = 0$ (here, 0 stands for the zero element of Z/Zp, which is actually the coset Zp). Since Z/Zp is a field, the product of non-zero elements in Z/Zp is never 0. Therefore, if a and b are integers that are not divisible by p, then the product ab is also not divisible by p. In other words, *a product is divisible by a prime number only if one of its factors is divisible by that prime number.*

Using this, one easily establishes the result on the unique factorization of positive integers (i.e., natural numbers) into products of prime numbers. The crude part of this result says that every natural number greater than 1 is a product of prime numbers. The following proof of this is based on the fact that every non-void set of natural numbers has a smallest element. Indeed, if S is such a set then $\bigcap(S)$ *is* the smallest element of S!

Now let us suppose that the above statement about factorization is false and derive a contradiction from this assumption. Let S be the set of all natural numbers greater than 1 that are *not* products of prime numbers. By our assumption, S is non-void. Accordingly, let q be the smallest element of S. By the definition of S, this number q cannot be a prime number. Therefore, $q = uv$, where u and v are natural numbers greater than 1. Since each of u and v is smaller than q, our choice of q implies that each is a product of prime numbers, whence we have the contradiction that their product q is a product of prime numbers.

The unicity of factorization says that if

$$p_1 \cdots p_m = q_1 \cdots q_n$$

where the p_i's and q_j's are prime numbers then $m = n$ and there is a bijective map from the index set $(1, \ldots, m)$ to itself, sending each i to i' say, such that $p_i = q_{i'}$ for each i. This is proved by induction on $n + m$, noting that p_1 must divide some q_j, and hence coincide with it, and then reducing the situation to a lower case by cancelling out p_1.

If $p_1, \ldots, p_m$ are prime numbers, then every prime factor of $1 + p_1 \cdots p_m$ must evidently be different from each p_i. Therefore, the crude part of the factorization result implies that *the set of all prime numbers is infinite.*

7. Suppose that R is a ring. There are many ways of using R for constructing more highly structured rings. The simplest such construction produces a ring whose elements are the maps from some fixed set S to R, and whose addition and multiplication are the *value-wise* operations. This means that, for maps f and g from S to R and elements s of S, one has

$$(f + g)(s) = f(s) + g(s) \quad \text{and} \quad (fg)(s) = f(s)g(s)$$

We wish to discuss a more subtle construction, in which the group, with addition as the group operation, of the new ring is the same as the above, but where the multiplication involves a given monoid structure of S, as follows.

We suppose that S is a monoid, and we indicate the composition of S by juxtaposition. We assume that the monoid composition of S is *of finite type*, in the sense that, for every element x of S, the set of elements (y, z) of $S \times S$

for which $yz = x$ is finite. Now we define the *convolution product* $f * g$ of the maps f and g from S to R by

$$(f * g)(x) = \sum_{yz=x} f(y)g(z)$$

where the expression on the right is the sum of the products $f(y)g(z)$ in R extended over the set of all pairs (y, z) with $yz = x$. The verification that this multiplication makes the group of maps from S to R into a ring involves no difficulties. If e is the neutral element of S then the neutral element for the multiplication of our new ring is the *characteristic function* of e, whose value at e is the neutral element 1 of the multiplication of R and whose values at all other elements of S are 0. We denote our new ring by $R[S]$. It is convenient to regard R as a subring of $R[S]$ by identifying each element r of R with the function from S to R whose value at e is r and whose values at all other elements of S are 0.

An element u of a ring A is called a *unit* if it has a reciprocal u^{-1} with respect to multiplication, so that $u^{-1}u = 1 = uu^{-1}$. It is easy to see that the units constitute a group whose composition map is the restriction of the multiplication of A and whose neutral element is the neutral element 1 of the multiplication of A. This group is called the *group of units* of A, and we shall denote it by $U(A)$.

Now let N^* denote the monoid of the positive natural numbers, with the multiplication as the composition map. Clearly, this monoid is of finite type in the above sense, so that we can construct the ring $Z[N^*]$, with the convolution product. We claim that *an element f of $Z[N^*]$ is a unit if and only if $f(1)$ is one of the two units* 1 *or* -1 *of* Z. This is evidently a necessary condition, because

$$1 = (f^{-1} * f)(1) = f^{-1}(1)f(1)$$

Conversely, suppose that $f(1) = \pm 1$. Then we define $f^{-1}(1) = f(1)$ and then extend the definition inductively as follows. Suppose that $f^{-1}(k)$ has already been defined for all natural numbers k in the range from 1 to some $n \geqq 1$. Then we define

$$f^{-1}(n + 1) = -f(1) \sum_{\substack{rs=n+1;\\ s>1}} f^{-1}(r)f(s)$$

This makes $(f^{-1} * f)(1) = 1$ and $(f^{-1} * f)(n) = 0$ whenever $n > 1$, so that f^{-1} is reciprocal to f.

The units of $Z[N^*]$ that are of particular interest in number theory are the so-called *multiplicative functions*, i.e., the functions f such that $f(1) = 1$ and $f(uv) = f(u)f(v)$ whenever $d(u, v) = 1$. If f and g are multiplicative, so are $f * g$ and f^{-1}, so that *the multiplicative functions constitute a subgroup of the group of units of* $Z[N^*]$. In order to show that $f * g$ is multiplicative, one notes that if $d(u, v) = 1$ and $xy = uv$ then $x = x_1x_2$ and $y = y_1y_2$, where $x_1y_1 = u$,

$x_2y_2 = v$ and $d(x_1, y_1) = 1 = d(x_2, y_2)$. In order to show that f^{-1} is multiplicative, one begins by defining a function g on the set of all prime powers such that $g(p^e) = f^{-1}(p^e)$ for every prime number p and every non-negative exponent e. Then one extends g to the full domain N^* simply by enforcing the multiplicative property. Now, by the first part, the product $f * g$ is multiplicative. By construction, it agrees with the neutral element of the group of units at each prime power p^e. Since both $f * g$ and this neutral element are multiplicative, it follows that $g = f^{-1}$, showing that f^{-1} is multiplicative.

In particular, let us consider the constant function on N^* with value 1. Thus, denoting this function by γ, we have $\gamma(n) = 1$ for every element n of N^*. We shall determine the reciprocal γ^{-1}. Since γ is multiplicative, so is γ^{-1}. Therefore, it suffices to find the values of γ^{-1} at the prime powers p^e. We must have $\gamma^{-1}(1) = 1$ and $(\gamma^{-1} * \gamma)(p) = 0$, i.e., $\gamma^{-1}(p)\gamma(1) + \gamma^{-1}(1)\gamma(p) = 0$, which forces $\gamma^{-1}(p) = -1$ for every prime number p. For each exponent $e > 1$, we have

$$0 = (\gamma^{-1} * \gamma)(p^e) = \gamma^{-1}(1) + \gamma^{-1}(p) + \cdots + \gamma^{-1}(p^e)$$

It follows that $\gamma^{-1}(p^e) = 0$ for every prime number p and every exponent $e > 1$. Hence, if $p_1, \ldots, p_k$ are pairwise distinct prime numbers and $e_1, \ldots, e_k$ are positive exponents, we have

$$\gamma^{-1}(p_1^{e_1} \cdots p_k^{e_k}) = \begin{cases} (-1)^k & \text{if each } e_i = 1 \\ 0 & \text{in all other cases} \end{cases}$$

This function γ^{-1} is called the *Moebius function*. It is usually denoted by μ.

For every element η of $Z[N^*]$, let us define the elements η' and η° of $Z[N^*]$ by

$$\eta'(x) = \sum_{yz = x} \mu(y)\eta(z)$$

$$\eta^\circ(x) = \sum_{y \mid x} \eta(y)$$

where $y \mid x$ is the customary abbreviation for "y divides x". Then the *Moebius inversion* formulae say that $(\eta')^\circ = \eta = (\eta^\circ)'$. These are immediate consequences of the fact that $\mu = \gamma^{-1}$, because $\eta' = \gamma^{-1} * \eta$ and $\eta^\circ = \gamma * \eta$.

Another important multiplicative function is *Euler's function*. This is usually denoted by φ, and the definition is that $\varphi(n)$ is the number of units of the ring Z/Zn. If $n = uv$, with $d(u, v) = 1$, then we find from Section 4 that there is a group isomorphism ρ from Z/Zn to $(Z/Zu) \times (Z/Zv)$, because $(Zu) \cap (Zv) = Zn$ and $Zu + Zv = Z$. If we define a multiplication for $(Z/Zu) \times (Z/Zv)$ by the formula $(x_1, y_1)(x_2, y_2) = (x_1x_2, y_1y_2)$ then this group becomes a ring, and our group isomorphism ρ is now a ring isomorphism. It follows directly from this that $\varphi(n) = \varphi(u)\varphi(v)$. Thus, φ is multiplicative.

Now observe that an element $t + Zn$ is a unit of Z/Zn if and only if $d(t, n) = 1$. If n is a prime power p^e, with positive exponent e, we see from this that the number of units of Z/Zp^e is equal to the total number p^e of elements minus the number p^{e-1} of multiples of the element $p + Zp^e$. This says that $\varphi(p^e) = p^e - p^{e-1}$.

Let us write ε for the identity function on N^*, so that $\varepsilon(n) = n$ for every n. We claim that $\gamma * \varphi = \varepsilon$. By the definition of γ, we have $(\gamma * \varphi)(n) = \sum_{u|n} \varphi\left(\frac{n}{u}\right)$ for every element n of N^*. For each divisor u of n, let S_u denote the set of natural numbers v satisfying $1 \leqq v \leqq n$ and $d(v, n) = u$. Evidently, these sets S_u are pairwise disjoint, and the union of the family of S_u's is the set $\{1, \ldots, n\}$. Hence, $\sum_{u|n} \mathbf{N}(S_u) = n$. On the other hand, a natural number v belongs to S_u if and only if $v = uw$, with $1 \leqq w \leqq \frac{n}{u}$ and $d\left(w, \frac{n}{u}\right) = 1$. Therefore, we have $\mathbf{N}(S_u) = \varphi\left(\frac{n}{u}\right)$. This establishes our claim that $\gamma * \varphi = \varepsilon$.

Multiplying by μ, we find that $\varphi = \mu * \varepsilon$. Explicitly, this says that we have $\varphi(n) = \sum_{u|n} \mu(u) \frac{n}{u}$ for every positive natural number n.

Exercises

1. Let R be a rectangle whose sides are of lengths a and b, where a and b are positive integers, with $a \geqq b$. If $a > b$, cut off a square of side length b from R by a cut parallel to the side of length b. Repeat this process with the remaining rectangle, etc. Show that, eventually, one is left with a square, which is the largest square with which R can be paved. Argue directly so as to conclude that the side length of this square is $d(a, b)$, and that the square tiles of this size needed for paving R can be assembled to form a rectangle whose sides have lengths $d(a, b)$ and $m(a, b)$, whence one recovers the result that $d(a, b)m(a, b) = ab$.

2. For arbitrary positive integers a and b, show that the number of group homomorphisms from Z/Za to Z/Zb is equal to $d(a, b)$.

3. Let σ be the surjective group homomorphism from $(Z/Za) \times (Z/Zb)$ to $Z/Zd(a, b)$ discussed at the end of Section 4. Write $d(a, b) = pa + qb$, $a = rd(a, b)$ and $b = sd(a, b)$, where p, q, r, s are appropriate integers. Now define a map η from Z to $(Z/Za) \times (Z/Zb)$ by the formula

$$\eta(x) = (-prx + Za, qsx + Zb)$$

Verify that η is a group homomorphism whose kernel contains $Zd(a, b)$, so that it defines a group homomorphism, η' say, from $Z/Zd(a, b)$ to $(Z/Za) \times (Z/Zb)$ in the natural way. Next, show that $\sigma \circ \eta'$ is the identity map on $Z/Zd(a, b)$. Using this, conclude that $(Z/Za) \times (Z/Zb)$ is isomorphic with (i.e., has a group isomorphism to) $(Z/Zm(a, b)) \times (Z/Zd(a, b))$.

4. For any group G, a group isomorphism from G to G is called an *automorphism* of G. The ordinary composition of maps makes the set of automorphisms of G into a group. If A is a commutative group, then the group of units of the ring $\mathrm{End}(A)$ is precisely the group of automorphisms of A. For an arbitrary positive integer a, determine the group of automorphisms of Z/Za. In particular, show that its cardinality is $\varphi(a)$, where φ is Euler's function.

5. In the notation of Section 7, note that $\gamma * \varepsilon$ is the sum of divisors function σ, defined by $\sigma(n) = \sum_{u|n} u$, and deduce that $\sum_{u|n} \mu(u)\sigma\left(\frac{n}{u}\right) = n$ for every positive integer n. Next, note that $\gamma * \gamma$ is the function τ whose value at each positive integer n is the number of positive divisors of n, and deduce that $\tau * \varphi = \sigma$. Finally, define the function α on N^* by $\alpha(n) = \mu(n)\tau(n)$. Using that $\gamma * \alpha$ is multiplicative, verify that $\sum_{u|n} \mu(u)\tau(u) = (-1)^q$, where q is the number of (distinct) prime numbers dividing n.

6. The notation being that of Section 7, determine ε^{-1} and φ^{-1}.

7. Let R be an arbitrary ring, and let N^+ stand for the monoid of the natural numbers, with addition as the composition map. Consider the ring $R[N^+]$. Let x denote the characteristic function of the element 1 of N^+, regarding x as an element of $R[N^+]$. By considering the powers x^n, with respect to the convolution product as the multiplication of $R[N^+]$, show that $R[N^+]$ appears as the "ring of formal power series in x with coefficients in R". Continue this interpretation by showing that the subring of $R[N^+]$ consisting of the functions taking *non-zero* values only on *finite* subsets of N^+ appears as the "ring of polynomials in x with coefficients in R".

Projects

1. Let n and s be positive integers, with $n \geqq s$. A *partition* of n into s summands is a sequence of s integers $1 \leqq a_0 \leqq \cdots \leqq a_{s-1}$ such that $\sum_{i=0}^{s-1} a_i = n$. Devise a computer program for producing all these partitions.

For this purpose, it is convenient to use the lexicographic ordering for the partitions. The main routine will take a partition as input and produce the lexicographically immediately preceding partition as output. Write $n = qs + r$, where q and r are non-negative integers and $r < s$. Observe that the lexicographically highest partition is given by $a_i = q$ for $i < s - r$ and $a_i = q + 1$ for $i \geqq s - r$. Denote this highest partition by $P(n, s)$. Now suppose $(a_0, \ldots, a_{s-1})$ is given as an input. If $a_i = 1$ for each $i < s - 1$ then this is the lowest partition, and the routine simply returns. Otherwise, let j be the largest index such that $a_i < a_j$ for each $i < j$. Then the immediately preceding partition is obtained by retaining each a_i for $i < j$, replacing a_j with $a_j - 1$ and following this up with $P(S_j + 1, s - j - 1)$, where $S_j = \sum_{i=j+1}^{s-1} a_i$.

2. Combine Project 1 above with Project 1 of Chapter I into a computer program that takes a finite set S of labels as input and produces all possible presentations of S as a union of a finite family of pairwise disjoint non-void subsets. There should be no duplications, with the agreement that presentations differing only in the ordering of the family of subsets or the internal ordering of individual subsets are regarded as identical.

3. Make a computer program effecting the multiplication in $Z[N^*]$, i.e., producing $(f * g)(n)$ whenever n and values of f and g at $1, \ldots, n$ have been given.

CHAPTER III

Fractions

1. The ring Z of integers is enlarged so as to become the field $\mathbf{Q}$ of rational numbers by introducing *fractions*. Technically, a fraction is a non-void subset S of $Z \times Z$ having the following properties.

(1) For every element (a, b) of S, the second entry b (the denominator) is different from 0.
(2) If (u, v) and (x, y) are elements of S then $uy = xv$.
(3) If (u, v) is an element of S and if x and y are integers such that $y \neq 0$ and $uy = xv$ then (x, y) is an element of S.

If (a, b) is any element of a fraction S, one writes $\frac{a}{b}$ (or a/b) for S. Thus, in the notation of (2), $\frac{u}{v} = S = \frac{x}{y}$.

One verifies directly, that *there is one and only one way of making the set of all fractions into a field* $\mathbf{Q}$ *such that the map sending each integer* x *to the fraction* $\frac{x}{1}$ *is an injective ring homomorphism.* By means of this map, we identify the integers x with the corresponding fractions $\frac{x}{1}$, so that it becomes legitimate to write $x = \frac{x}{1}$.

For computational purposes, fractions are frequently represented by sequences of digits. In precise terms, this formalism is as follows. Let b be an integer greater than 1. We denote the set of integers $\{0, 1, \ldots, b - 1\}$ by $R(b)$. The members of $R(b)$ serve as the digits. For the present discussion, we confine our attention to non-negative fractions less than 1. Accordingly, let

u and v be non-negative integers with $u < v$, and consider the fraction $\frac{u}{v}$. The digital representation of this fraction is a sequence $(d_1, d_2, \ldots)$, where each d_i belongs to $R(b)$ and

$$d_1b^{-1} + \cdots + d_ib^{-i} \leqq \frac{u}{v} < d_1b^{-1} + \cdots + d_ib^{-i} + b^{-i}$$

for each index i.

In order to make the determination of the digits d_i from these conditions explicit, we define the *integer part* $[\rho]$ of a number ρ as the largest integer r such that $r \leqq \rho$. Then the above d_i's are given by the following recursion, where $i = 1, 2, \ldots$.

$$f_0 = \frac{u}{v}$$

$$d_i = [bf_{i-1}]$$

$$f_i = bf_{i-1} - d_i$$

Using that vf_0 is an integer, we see from this recursion that vf_i is an integer for every i. Also, it is clear that $0 \leqq f_i < 1$, so that $f_i = r_i/v$, where r_i is an element of $R(v)$. Thus, our recursion may be written so as to involve only integers, as follows.

$$r_0 = u$$

$$d_i = [br_{i-1}/v]$$

$$r_i = br_{i-1} - vd_i$$

If some r_i equals 0 then the sequence of d_j's consists of 0's from $j = i + 1$ onwards. Otherwise, each r_i belongs to the set $R(v)\setminus\{0\}$ of cardinality $v - 1$, whence we see that there must be indices i and j such that $i < j < v$ and $r_j = r_i$. Then we have $d_{j+1} = d_{i+1}$ and hence $r_{j+1} = r_{i+1}$. Let p denote the smallest such index i, and let t be the smallest positive integer for which $r_{p+t} = r_p$. Then our sequence of d_i's is $(d_1, \ldots, d_p, e_1, \ldots, e_t, e_1, \ldots, e_t, \ldots)$, where $t < v$ and $e_k = d_{p+k}$. In other words, the sequence of digits is periodic immediately beyond the pth term, and $(e_1, \ldots, e_t)$ is the period, with $t < v$.

We observe that the case where $t = 1$ and $e_1 = b - 1$ does *not* arise as above from a fraction $\frac{u}{v}$. Indeed, the rational numbers corresponding to the finite initial sections of our sequence in that case are smaller than $b^{-p} + \sum_{i=1}^{p} d_ib^{-i}$, but eventually exceed every rational number smaller than this sum. Consequently, the fraction represented by the sequence is equal to $b^{-p} + \sum_{i=1}^{p} d_ib^{-i}$, and the recursion for the d_j's would yield $d_j = 0$ from $j = p + 1$ onwards, at the latest. However, in every other case, an eventually periodic sequence is, in fact, the sequence obtained as above from a fraction $\frac{u}{v}$.

Explicitly, if $e = e_1 b^{-1} + \cdots + e_t b^{-t}$, then this fraction is equal to

$$d_1 b^{-1} + \cdots + d_p b^{-p} + b^{-p} e/(1 - b^{-t})$$

as is seen by considering the sums $\sum_{i=0}^{n} eb^{-it}$ with $n = 0, 1, \ldots$.

The indices p and t have the following arithmetic interpretations. First, note that the rational numbers that can be written in the form $\frac{x}{y}$, where x is an integer and y is a positive integer *prime to* b, in the sense that $d(y, b) = 1$, constitute a subring of $\mathbf{Q}$. Let us denote this subring by $\mathbf{Q}_b$. Clearly, if ρ is any rational number, there are integer exponents q such that ρb^q belongs to $\mathbf{Q}_b$. If $\rho \neq 0$, then there is a smallest such exponent (which, of course, may be negative). Let us call this smallest q the *b-order* of ρ.

Now we return to our digital representation of $\frac{u}{v}$. We had already assumed that this representation does not end with a period of length 1 consisting of the digit 0. In the case where $p = 0$, so that $\frac{u}{v} = eb^t/(b^t - 1)$, let us assume that $e_t \neq 0$. We claim that, under these assumptions, *the index p is precisely the b-order of $\frac{u}{v}$.*

In order to prove this, we note that $\frac{u}{v} b^p$ can be written as a fraction with denominator $b^t - 1$, showing that the b-order of $\frac{u}{v}$ is at most equal to p. Now suppose that the b-order of $\frac{u}{v}$ is actually smaller than p. Then we have $\frac{u}{v} b^{p-1} = \frac{x}{y}$, with $d(y, b) = 1$. This yields the equation

$$\{(b^t - 1)(d_1 b^{p-1} + \cdots + d_p) + eb^t\} y = xb(b^t - 1)$$

Together with the fact that $d(y, b) = 1$, this implies that $-d_p + e_t$ is divisible by b. In the case where $p = 0$, we must interpret this to mean that e_t is divisible by b. By our extra assumption, we have $e_t \neq 0$ in this case and, since e_t belongs to $R(b)$, it cannot be divisible by b. If $p \neq 0$, then the condition that $-d_p + e_t$ be divisible by b implies $d_p = e_t$. However, this contradicts our assumption that p is chosen so as to be as small as possible, because the digit representation starts with $d_1, \ldots, d_{p-1}$ and is followed by the period $(e_t, e_1, \ldots, e_{t-1})$. Thus, in each case we have allowed, the assumption that the b-order of $\frac{u}{v}$ is smaller than p leads to a contradiction. This completes the proof of our assertion concerning the index p.

The case where $p = 0$ may be completed by the observation that the b-order of $eb^t/(b^t - 1)$ equals $s - t$, where s is the largest index with $e_s \neq 0$.

Recalling that the case where $t = 1$ and $e_1 = b - 1$ does not arise as above from a fraction $\frac{u}{v}$, we see that the periodic part of the digital representation of $\frac{u}{v}$ is equal to

$$\frac{u}{v} - b^{-p}\left[\frac{u}{v}b^p\right]$$

where the square brackets indicate the integer part function. Therefore, having characterized p, we loose nothing in confining our discussion of the index t to the case where $\frac{u}{v} = eb^t/(b^t - 1)$. Moreover, we shall assume that $\frac{u}{v}$ is the fully reduced form of our fraction, i.e., that $d(u, v) = 1$. The interpretation of t involves the group $U(Z/Zv)$ of units of the ring Z/Zv. Generally, if x is an element of a finite group, then the *order* of x is the cardinality of the smallest subgroup containing x. This is the smallest non-negative exponent s such that x^s is the neutral element.

We have $u(b^t - 1) = eb^tv$. Since eb^t is an integer and $d(u, v) = 1$, it follows that $b^t - 1$ belongs to Zv. Therefore, $b + Zv$ is a unit of Z/Zv and its order as an element of the group $U(Z/Zv)$ is no greater than t. If s denotes this order, we have $b^s - 1 = hv$, where h is a positive integer. Now

$$\frac{u}{v} = \frac{hu}{hv} = \frac{hu}{b^s - 1} = \frac{b^{-s}hu}{1 - b^{-s}}$$

Since hu is a positive integer, we may write

$$b^{-s}hu = m + f_1b^{-1} + \cdots + f_sb^{-s}$$

where m is a non-negative integer and each f_i belongs to $R(b)$. By our original assumption, $\frac{u}{v} < 1$, whence $m/(1 - b^{-s}) < 1$, so that $m = 0$. Therefore, the above expression for $\frac{u}{v}$ shows that the digital representation of $\frac{u}{v}$ is periodic, with period $(f_1, \ldots, f_s)$. Since t is the minimal period length, it follows that we must have $t \leqq s$. From the beginning of this argument, we know that $t \geqq s$. Thus, $t = s$. Our conclusion is that *the period length t is equal to the order of b + Zv as an element of the group of units of the ring Z/Zv.*

Finally, we have seen that *the positive fractions* <1 *having purely periodic digital representations with the last digit of the period not* 0 *are precisely those whose reduced forms are* $\frac{u}{v}$, *with* $0 < u < v$, $d(v, b) = 1$ *and u not divisible by b.*

The following simple example may serve to clarify the above discussion. Let $b = 6$, $u = 9$, $v = 10$. The b-order of 9/10 is 1, so that $p = 1$. The periodic part is therefore

$$\frac{9}{10} - \frac{1}{6}[(9/10)6] = \frac{1}{6}\frac{2}{5}$$

This shows that the period is the same as that of the purely periodic fraction $\frac{2}{5}$. The order of the unit $6 + Z5$ of $Z/Z5$ is equal to 1, so that the period length is 1. Actually, we have $d_1 = 5$ and $e_1 = 2$.

2. In order to proceed we must deal systematically with sequences of rational numbers. In precise terms, an infinite sequence of rational numbers is a map from N to $\mathbf{Q}$. With the value-wise operations, these sequences constitute a ring, which we shall denote by S. An element of S is called a *Cauchy sequence* (of rational numbers) if its values at sufficiently large elements of N differ from each other by arbitrarily little. Expressed rigorously, this means than *an element f of S is a Cauchy sequence if, for every positive rational number e, there is a natural number n_e such that $f(p)$ and $f(q)$ differ from each other by less than e whenever p and q are greater than n_e.*

One verifies directly that the Cauchy sequences constitute a subring of S, which we denote by T. Let us say that a Cauchy sequence f is *negligible* if it approaches 0 as a limit, i.e., if, for every positive rational number e, there is a natural number m_e such that $f(q)$ differs from 0 by less than e whenever q is greater than m_e. It is easy to verify that the negligible Cauchy sequences constitute an ideal of T. We denote this ideal by J.

We consider the factor ring T/J. First, we note that the map sending each rational number ρ to the coset containing the constant function on N with value ρ is an injective ring homomorphism from $\mathbf{Q}$ to T/J. Next, if f is a Cauchy sequence not belonging to J, there is a positive rational number ρ and a natural number n such that $|f(p)| > \rho$ whenever $p > n$. Choose a function g from N to $\mathbf{Q}$ such that $g(p) = f(p)^{-1}$ whenever $p > n$. It is easy to verify that g is a Cauchy sequence. Clearly, $fg - 1$ belongs to J, where 1 stands for the constant function on N whose value is the rational number 1. This means that the canonical image of g in T/J is reciprocal to the canonical image of f. Thus, T/J is a field.

The above remark concerning the values of f needs to be refined only slightly in order to show that the notion of positivity in $\mathbf{Q}$ leads to a notion of positivity in T/J having the same compatibility with addition and multiplication, i.e., being such that sums and products of positive elements are positive, and having the property that, for every element α of T/J other than 0, exactly one of α or $-\alpha$ is positive. In precise terms, one defines α to be positive if there is a positive rational number ρ such that, for every Cauchy sequence f belonging to α, one has $f(n) > \rho$ for all sufficiently large natural numbers n.

In the usual way, this notion of positivity yields the definition of an ordering of T/J, such that our injective map from $\mathbf{Q}$ to T/J is order preserving and the image of $\mathbf{Q}$ is *dense* in T/J, in the sense that, for every element α of T/J and every positive rational number e, there is a rational number whose image in T/J differs from α by less than e. When equipped with this structure, T/J is the *field* $\mathbf{R}$ *of real numbers*. By means of the map defined above, *the field* $\mathbf{Q}$ *of rational numbers is identified with a dense subfield of the field* $\mathbf{R}$ *of real numbers*.

The gain derived from constructing $\mathbf{R}$ consists in the basic fact that *every Cauchy sequence in* $\mathbf{R}$ *converges to a limit in* $\mathbf{R}$. This means that, if γ is a Cauchy sequence with values $\gamma(n)$ in $\mathbf{R}$, there is an element c in $\mathbf{R}$ such that, for every positive rational number e, there is a natural number n_e such that $\gamma(p)$ differs from c by less than e whenever $p > n_e$. In order to see this, we construct a Cauchy sequence as follows. First, for every n in N, we choose a Cauchy sequence τ_n representing $\gamma(n)$, in the sense that $\tau_n + J = \gamma(n)$. Now, using that τ_n is a Cauchy sequence, we choose a natural number n' such that $\tau_n(p)$ differs from $\tau_n(q)$ by less than $1/n$ whenever both $p \geqq n'$ and $q \geqq n'$. Finally, let σ be the element of S defined by $\sigma(n) = \tau_n(n')$ for every n in N. It is easy to verify that σ belongs to T and that the given Cauchy sequence γ in $\mathbf{R}$ converges to the limit $\sigma + J$.

3. Let $\rho_0, \rho_1, \ldots, \rho_n$ be real numbers, with $\rho_i > 0$ for each index i other than 0. We define the real number $]\rho_0, \ldots, \rho_n[$ by the following recursion.

$$]\rho_0[\ = \rho_0, \qquad]\rho_0, \ldots, \rho_n[\ = \rho_0 + \frac{1}{]\rho_1, \ldots, \rho_n[}$$

It is readily verified from this recursion that, if ρ is any positive real number, one has

$$]\rho_0, \ldots, \rho_n, \rho[\ = \ ]\rho_0, \ldots, \rho_{n-1}, \rho_n + \rho^{-1}[$$

where the expression on the right is to be interpreted as $\rho_0 + \rho^{-1}$ when $n = 0$.

Now let $(a_0, a_1, \ldots)$ be an infinite sequence of integers, with $a_i > 0$ for every index i other than 0. Then the sequence of rational numbers whose nth term is $]a_0, \ldots, a_n[$ for every n in N is called the *simple continued fraction* defined by $(a_0, a_1, \ldots)$. In order to investigate such a continued fraction, one introduces two auxiliary sequences of integers, as follows.

$$h_{-2} = 0, h_{-1} = 1, h_i = a_i h_{i-1} + h_{i-2} \quad \text{for } i \geqq 0;$$

$$k_{-2} = 1, k_{-1} = 0, k_i = a_i k_{i-1} + k_{i-2} \quad \text{for } i \geqq 0.$$

First, we note that $k_n > 0$ whenever $n \geqq 0$. Next, using the formal property of $]\cdots[$ noted above, one verifies by induction that, if ρ is any positive real number and $n \geqq 0$, then

$$]a_0, \ldots, a_{n-1}, \rho[\ = (\rho h_{n-1} + h_{n-2})/(\rho k_{n-1} + k_{n-2})$$

In particular, this gives $]a_0, \ldots, a_n[\ = h_n/k_n$.

An induction shows also that, for every non-negative n, one has

$$h_{n-1}k_{n-2} - h_{n-2}k_{n-1} = (-1)^n \quad \text{and}$$
$$h_n k_{n-2} - h_{n-2}k_n = (-1)^n\, a_n$$

Writing r_n for $]a_0, \ldots, a_n[$, we may express the last result, for $n \geqq 2$, in the form

$$r_n - r_{n-2} = (-1)^n\, a_n/(k_n k_{n-2})$$

This shows that, for all $n \geqq 0$, we have

$$r_{2n} < r_{2(n+1)} \quad \text{and} \quad r_{2(n+1)+1} < r_{2n+1}$$

From the first of the above relations between the h's and k's, we see that, for every $n \geqq 2$,

$$r_{n-1} - r_{n-2} = (-1)^n/(k_{n-1}k_{n-2})$$

which shows that $r_{2n+1} > r_{2n}$ for every $n \geqq 0$. Putting our inequalities for the r's together, we find that, for all positive indices n and m, we have

$$r_{2n} < r_{2(n+m)} < r_{2(n+m)+1} < r_{2m+1}$$

We may summarize these results as follows. Let $u_n =]a_0, \ldots, a_{2n}[$ and $v_n =]a_0, \ldots, a_{2n+1}[$. Then *the sequence with terms u_n is strictly increasing, the sequence with terms v_n is strictly decreasing, and each u_n is smaller than each v_m.*

Now observe that the k_n's constitute a strictly increasing sequence of positive integers from $n = 0$ onwards, so that our above expression for $r_{n-1} - r_{n-2}$ shows that $v_n - u_n$ approaches the limit 0 as n increases without bound. Together with our summary, this shows that *the continued fraction defined by $(a_0, a_1, \ldots)$ is a Cauchy sequence and therefore approaches a certain real number α as a limit.* In the customary notation for limits, we have $\alpha = \lim_{n\to\infty}(]a_0, \ldots, a_n[)$.

We show that *α is not a rational number.* From our above summary and the expression for $r_{n-1} - r_{n-2}$, it is clear that

$$0 < |\alpha - r_n| < |r_{n+1} - r_n| = (k_n k_{n+1})^{-1}$$

From this, we obtain

$$|k_n\alpha - h_n| < 1/k_{n+1}$$

for all $n \geqq 0$. Now suppose that, contrary to what we wish to show, $\alpha = u/v$, where u and v are integers and $v > 0$. Then the last inequality above gives

$$|k_n u - h_n v| < v/k_{n+1}$$

Since the expression on the left is an integer, while k_{n+1} grows without bound with n, this forces $r_n = u/v$ for all sufficiently large indices n. On the other hand, we have seen above that $r_{n-1} - r_{n-2}$ is never equal to 0. Thus, our assumption that α is rational has led to a contradiction.

Next, we observe that *the limit* α *determines the sequence* $(a_0, a_1, \ldots)$. Indeed, we know that $r_0 < \alpha < r_1$, i.e., $a_0 < \alpha < a_0 + a_1^{-1}$. Since a_0 is an integer and $a_1 \geqq 1$, this implies that $a_0 = [\alpha]$. From the recursion defining $]a_0, \ldots, a_n[$, we see that the sequence with terms $]a_1, \ldots, a_n[$ approaches the limit $(\alpha - [\alpha])^{-1}$ as n increases without bound. Therefore, repetition of our argument shows that we must have $a_1 = [1/(\alpha - [\alpha])]$, etc. In this way, each a_i is seen to be determined by α.

Our last argument suggests a proof of the fact that *every irrational real number is the limit of a simple continued fraction.* Given an irrational real number α, we define a sequence of such numbers as follows. We begin with $\alpha_0 = \alpha$. Having defined an irrational real number α_i for some index i, we note that $\alpha_i - [\alpha_i]$ is irrational, and we define $\alpha_{i+1} = (\alpha_i - [\alpha_i])^{-1}$. Now we verify that the simple continued fraction constructed from the sequence of integers $a_i = [\alpha_i]$ has the limit α. Using the identity

$$]\rho_0, \ldots, \rho_n, \rho[\;=\;]\rho_0, \ldots, \rho_{n-1}, \rho_n + \rho^{-1}[$$

in an induction, one shows that

$$\alpha_0 =]a_0, \ldots, a_{n-1}, \alpha_n[$$

for every positive index n. Using the auxiliary sequences of h's and k's, this result can be put in the form

$$\alpha_0 = (\alpha_n h_{n-1} + h_{n-2})/(\alpha_n k_{n-1} + k_{n-2})$$

from which one deduces that, for every positive index n,

$$\alpha_0 - r_{n-1} = \frac{h_{n-2}k_{n-1} - h_{n-1}k_{n-2}}{(\alpha_n k_{n-1} + k_{n-2})k_{n-1}} = \frac{(-1)^{n-1}}{(\alpha_n k_{n-1} + k_{n-2})k_{n-1}}$$

Since $\alpha_n > 0$ and the k_i's constitute a strictly increasing sequence of positive integers (for $i \geqq 0$), this shows that the sequence with terms r_n approaches the limit α_0 as n grows without bound. This means that the given irrational number α is indeed the limit of our continued fraction.

A notable feature of the approximation of α *by the fractions* $r_n = h_n/k_n$ *is that these fractions are reduced*, i.e., $d(h_n, k_n) = 1$, *and* h_n/k_n *differs from* α *by less than* k_n^{-2}, as is visible directly from the above results.

4. Let us say that a real number α is *quadratic* if it is irrational, but is a root of a polynomial of degree 2 with rational coefficients, so that there are rational numbers s and t such that $\alpha^2 + s\alpha + t = 0$. Every such number α can be written in the form $(u + \sqrt{d})/v$, where u, v and d are integers, $v \neq 0$ and d is positive but not the square of an integer (as usual, $\sqrt{d}$ denotes the positive real number whose square is d). Conversely, every number of this form is a quadratic number.

We shall show that *the quadratic numbers are precisely those irrational real numbers whose continued fraction representations are eventually periodic.*

Consider a quadratic number α, written as above. Multiplying numerator and denominator by $|v|$, we rewrite α in one of the forms

$$(\pm uv + \sqrt{dv^2})/\pm v^2.$$

Writing d for dv^2, etc., we have thus obtained $\alpha = (m_0 + \sqrt{d})/q_0$, where now the triple (m_0, q_0, d) satisfies the conditions made above for (u, v, d) but where, furthermore, q_0 divides $d - m_0^2$. Accordingly, we write $d - m_0^2 = t_0 q_0$. Now, in the notation used in Section 3 for the construction of the continued fraction representation of α, we obtain

$$\alpha_1 = (\alpha_0 - a_0)^{-1} = \frac{q_0}{m_0 + \sqrt{d} - q_0 a_0} = \frac{q_0(\sqrt{d} - m_0 + q_0 a_0)}{d - (m_0 - q_0 a_0)^2}$$

which simplifies to

$$\alpha_1 = \frac{(a_0 q_0 - m_0) + \sqrt{d}}{t_0 + 2m_0 a_0 - q_0 a_0^2}$$

Write m_1 for $a_0 q_0 - m_0$ and q_1 for $t_0 + 2m_0 a_0 - q_0 a_0^2$. Then we have

$$\alpha_1 = (m_1 + \sqrt{d})/q_1 \quad \text{and} \quad d - m_1^2 = q_0 q_1$$

The second equation shows that the triple (m_1, q_1, d) satisfies the same conditions as the triple (m_0, q_0, d). Therefore, we can continue this construction so as to define a sequence of integers m_i and a sequence of non-zero integers q_i such that, for all $i \geqq 0$,

$$\begin{aligned} \alpha_i &= (m_i + \sqrt{d})/q_i \\ m_{i+1} &= [\alpha_i]q_i - m_i \\ q_{i+1} &= (d - m_{i+1}^2)/q_i \end{aligned}$$

We know from Section 3 that the continued fraction representation of α is given by the sequence of integers $a_i = [\alpha_i]$.

We are working in the subfield, D say, of $\mathbf{R}$ consisting of the numbers of the form $u + v\sqrt{d}$, where u and v range over $\mathbf{Q}$. It is easy to verify that the map δ from D to D defined by $\delta(u + v\sqrt{d}) = u - v\sqrt{d}$ is a ring homomorphism. Recall from Section 3 that, for all positive indices n,

$$\alpha_0 = (\alpha_n h_{n-1} + h_{n-2})/(\alpha_n k_{n-1} + k_{n-2})$$

Applying the homomorphism δ to this, we obtain

$$\delta(\alpha_0) = (\delta(\alpha_n)h_{n-1} + h_{n-2})/(\delta(\alpha_n)k_{n-1} + k_{n-2})$$

which gives

$$\delta(\alpha_n) = (h_{n-2} - k_{n-2}\delta(\alpha_0))/(k_{n-1}\delta(\alpha_0) - h_{n-1})$$
$$= -\frac{k_{n-2}}{k_{n-1}}\frac{r_{n-2} - \delta(\alpha_0)}{r_{n-1} - \delta(\alpha_0)}$$

We know that, as n grows without bound, r_n approaches the limit α_0. Since $\alpha_0 - \delta(\alpha_0) \neq 0$, we may conclude from this that $(r_{n-2} - \delta(\alpha_0))/(r_{n-1} - \delta(\alpha_0))$ approaches the limit 1 as n grows without bound. Together with this, our above expression for $\delta(\alpha_n)$ shows that $\delta(\alpha_n) < 0$ for all sufficiently large indices n, say for all $n > n_1$. On the other hand, the definition of α_n shows that $\alpha_n > 0$ for every n other than 0. Therefore, we have $\alpha_n - \delta(\alpha_n) > 0$ whenever $n > n_1$. This says that $2\sqrt{d}/q_n > 0$ and hence that $q_n > 0$ for all $n > n_1$.

Since $q_n q_{n+1} = d - m_{n+1}^2$, it follows that $|m_{n+1}| < \sqrt{d}$ whenever $n > n_1$. At the same time, we see that $q_n < d$. Thus, the m_i's and q_i's are bounded in absolute value. Consequently, there must be indices $j < k$ such that $m_j = m_k$ and $q_j = q_k$. This implies that $\alpha_j = \alpha_k$ and that the sequence of a_n's becomes periodic, with period $(a_j, \ldots, a_{k-1})$.

Conversely, suppose that $(a_0, a_1, \ldots)$ is an eventually periodic sequence of integers, with $a_n > 0$ for each positive index n, having a period $(a_j, \ldots, a_{k-1})$, where $j < k$. Let τ denote the limit of the simple continued fraction defined by the purely periodic part of the sequence of a_n's. We have seen in Section 3 that, if α_n denotes the limit of the continued fraction defined by the sequence $(a_n, a_{n+1}, \ldots)$, then $\alpha_0 =]a_0, \ldots, a_{n-1}, \alpha_n[$. Because of the periodicity, this yields $\tau =]a_j, \ldots, a_{k-1}, \tau[$. We know from Section 3 that the expression on the right can be written in the form $(\tau u + u')/\tau v + v')$, where u, u', v, v' are certain integers depending on $a_j, \ldots, a_{k-1}$. Given the fact that τ is not rational, this evidently implies that τ is a quadratic number. Finally, the continued fraction recursion shows that the limit of the full continued fraction defined by the given sequence $(a_0, a_1, \ldots)$ is therefore also a quadratic number.

5. We shall exhibit a classical result, due to Liouville, concerning approximations to roots of polynomials with rational coefficients by rational numbers. We consider polynomials $f(x) = \sum_{i=0}^{n} c_i x^i$, where the coefficients c_i are rational numbers, and we interpret such a polynomial as a map from **R** to **R** in the conventional way, denoting its value at a real number ρ by $f(\rho)$. We say that $f(x)$ is of degree n if $c_n \neq 0$. Suppose that f is of degree n and that τ is an irrational real number such that $f(\tau) = 0$. Liouville's theorem says that there is a positive real number ρ such that, for all integers u and v, with $v > 0$, one has $\left|\tau - \dfrac{u}{v}\right| \geqq \rho v^{-n}$.

First, let us reduce the situation to the case where f has no rational root. Suppose that ρ is a rational root of f. Using that

$$x^e - \rho^e = (x - \rho) \sum_{i=1}^{e} \rho^{e-i} x^{i-1}$$

for $e = 1, 2, \ldots$ and writing $f(x)$ as $f(x) - f(\rho)$, we find that

$$f(x) = (x - \rho)g(x),$$

where $g(x)$ is a polynomial of degree $n - 1$ with rational coefficients. If τ is as above then $g(\tau) = 0$ and, since $v^{-(n-1)} \geqq v^{-n}$, nothing is lost in replacing f with g. Repeating this argument, if necessary, we arrive eventually at the case where the relevant polynomial has no rational root.

In this case, we prepare the situation further by multiplying f with a suitable positive integer so as to achieve that all the coefficients become integers. Accordingly, we assume from now on that f is a polynomial of degree n with integer coefficients such that $f(\tau) = 0$, while f has no rational root.

For every real number σ, we may write

$$f(\sigma) = f(\sigma) - f(\tau) = (\sigma - \tau) \sum_{i=1}^{n} c_i \left(\sum_{q=1}^{i} \sigma^{i-q} \tau^{q-1} \right)$$

Estimating the sum on the right, it is easy to see from this that there is a positive real number ρ such that $\rho |f(\sigma)| \leqq |\sigma - \tau|$ for every real number σ such that $|\sigma - \tau| < 1$. As we evidently may, we choose $\rho \leqq 1$. Now let us take $\sigma = \frac{u}{v}$, where u and v are integers and $v > 0$. If $\left| \tau - \frac{u}{v} \right| \geqq 1$, then we have $\left| \tau - \frac{u}{v} \right| \geqq \rho v^{-n}$, because $\rho \leqq 1$. Otherwise, our choice of ρ gives

$$\left| \tau - \frac{u}{v} \right| \geqq \rho \left| f\left(\frac{u}{v}\right) \right|$$

Since the coefficients of f are integers, $v^n f\left(\frac{u}{v}\right)$ is an integer. Since f has no rational root, this integer is not 0. Hence, $\left| f\left(\frac{u}{v}\right) \right| \geqq v^{-n}$, so that

$$\left| \tau - \frac{u}{v} \right| \geqq \rho v^{-n}.$$

Exercises

1. Show that every ordinary fraction can be written in one and only one way as the sum of an integer and fractions of the form u/p^e, where p is a prime, e a positive exponent and u an integer of absolute value less than p.

2. Let A be the group of rational numbers, with addition as the group composition. Let M be the group of the non-zero rational numbers, with multiplication as the group composition. Show that the only group homomorphism from A to M is the *trivial* one, sending every element of A to 1 (use the prime factorization theory for N in looking at M).

3. Suppose that $(\alpha_0, \alpha_1, \ldots)$ is a sequence of real numbers such that $\alpha_n \leqq \alpha_{n+1}$ for every index n. Show that this is a Cauchy sequence if and only if there is a real number β such that $\alpha_n \leqq \beta$ for every index n.

4. Prove that every real number is the limit of a non-decreasing sequence of rational numbers.

5. Let ρ be an arbitrary rational number. Show that, for some index $n \geqq 0$, one has $\rho =]a_0, \ldots, a_n[$, where the a_i's are integers, with $a_i > 0$ whenever $i > 0$ (the a_i's can be obtained as quotients from a chain of divisions as used for finding a greatest common divisor).

6. Let b be an integer > 1, and let $\beta_n = \sum_{i=1}^{n+1} b^{-i!}$ for $n = 0, 1, \ldots$. Using the result of Section 5, show that the limit of the Cauchy sequence $(\beta_0, \beta_1, \ldots)$ is not a root of any (non-zero) polynomial with rational coefficients.

Projects

1. Devise a computer program for calculating the digital representation of an ordinary fraction. In the notation of Section 1, the program is to find $d_1, \ldots, d_p$ and the period $(e_1, \ldots, e_t)$. A method for discovering the period without storing a possibly long string of remainders r_i is as follows. For $i = 1, 2, \ldots,$ one tests for coincidence of r_{i-1} with r_{2i-1}. If the first coincidence is for $i = m$ and the second for $i = n$, then the periodicity t is $n - m$, and the periodic part starts with $e_1 = d_m$.

2. Consider quadratic numbers $(u + \sqrt{d})/v$, as in Section 4. In order to find the continued fraction representation of such a number, α say, one must calculate the integer parts $[\alpha_i]$ for the quadratic numbers $\alpha_0 = \alpha$ and the subsequent α_i's as described in Section 4. Devise a computer program accomplishing this with pure integer arithmetic, so that there are no rounding errors. The simplest case is $\alpha = \sqrt{d}$. Here, one finds the largest positive integer p satisfying $p^2 < d$ simply by starting with 1 and incrementing. The final such p is $[\sqrt{d}]$, of course. Generally, one distinguishes two cases, according to the sign of v. In the case where $v > 0$, show that $[(u + \sqrt{d})v]$ is either equal to $[(u + [\sqrt{d}])/v]$ or equal to this integer plus 1, and that the decision can be made by comparing the square of an appropriate integer with d. The case where $v < 0$ can be handled in a similar fashion.

CHAPTER IV

Linearity

1. Consider a commutative group V, whose composition map we indicate by $+$. Recall that the endomorphisms of V constitute a ring End(V). Now we suppose that there is given a ring homomorphism σ from some field F to End(V). The customary terminology for referring to this situation is to say that σ makes V into a *vector space* over F. For α in F and v in V, one abbreviates $\sigma(\alpha)(v)$ by αv, and one calls $\sigma(\alpha)$ the *scalar multiplication* by α. The fact that $\sigma(\alpha)$ is an endomorphism of V is expressed by the formula

$$\alpha(v_1 + v_2) = \alpha v_1 + \alpha v_2$$

and the fact that σ is a ring homomorphism is expressed by the formulas

$$(\alpha_1 + \alpha_2)v = \alpha_1 v + \alpha_2 v$$
$$(\alpha_1 \alpha_2)v = \alpha_1(\alpha_2 v)$$

Note that the field F is actually a vector space over F, the scalar multiplication being simply the multiplication of the field structure of F.

If $(v_1, \ldots, v_n)$ is a finite sequence of elements of V, then a *linear combination* of $(v_1, \ldots, v_n)$ is an element of V of the form $\alpha_1 v_1 + \cdots + \alpha_n v_n$, with each α_i in F. One says that such a sequence is *linearly dependent* if there are elements α_i in F, not all equal to the zero element of F, such that the above sum is the zero element of V. Equivalently, $(v_1, \ldots, v_n)$ is a linearly dependent sequence if, for at least one index i, the element v_i is a linear combination of the sequence obtained by deleting v_i from $(v_1, \ldots, v_n)$. We agree that a linear combination of the empty sequence is the zero element of V.

If the sequence $(v_1, \ldots, v_n)$ is not linearly dependent, then it is said to be *linearly independent*. In this case, the v_i's are pairwise distinct, so that we may refer to $\{v_1, \ldots, v_n\}$ also as a *linearly independent subset* of V. More generally, a subset T of V is called linearly independent if every finite subset of T is linearly independent.

Let S be a non-void set, and let V be the set of all maps from S to the field F. We make V into a vector space over F by means of the value-wise operations. Thus, if f and g are elements of V and α is an element of F, then $f + g$ and αf are defined by the formulas

$$(f + g)(s) = f(s) + g(s), \qquad (\alpha f)(s) = \alpha f(s)$$

The following simple fact is the basis for the technical control of linear independence.

Let $\{v_1, \ldots, v_n\}$ be a linearly independent subset of the vector space V of all maps from S to F. There are linear combinations $f_1, \ldots, f_n$ of $(v_1, \ldots, v_n)$ and corresponding elements $s_1, \ldots, s_n$ of S such that $f_i(s_i) = 1$ for each index i, while $f_i(s_j) = 0$ whenever $i \neq j$.

The proof of this begins with the remark that there is an element s_1 in S such that $v_1(s_1) \neq 0$. Set

$$g_{1,1} = v_1(s_1)^{-1}v_1$$

For each index i other than 1, set

$$g_{i,1} = v_i - v_i(s_1)g_{1,1}$$

Then $g_{1,1}, \ldots, g_{n,1}$ are linearly independent linear combinations of $(v_1, \ldots, v_n)$, and we have $g_{1,1}(s_1) = 1$ while $g_{i,1}(s_1) = 0$ for each index i other than 1.

Now suppose that, for some index $k < n$, we have already found linearly independent linear combinations $g_{1,k}, \ldots, g_{n,k}$ of $(v_1, \ldots, v_n)$ and corresponding elements $s_1, \ldots, s_k$ of S such that $g_{i,k}(s_j)$ is equal to 1 if $j = i$, and equal to 0 for the other indices j. Then there is an element s_{k+1} in S such that $g_{k+1,k}(s_{k+1}) \neq 0$. Now we put

$$g_{k+1,k+1} = g_{k+1,k}(s_{k+1})^{-1}g_{k+1,k}$$

and, for every index i other than $k + 1$,

$$g_{i,k+1} = g_{i,k} - g_{i,k}(s_{k+1})g_{k+1,k+1}$$

Evidently, this process can be continued until $k = n$, and if $f_i = g_{i,n}$ then $f_1, \ldots, f_n$ and $s_1, \ldots, s_n$ satisfy our requirements.

Note the implication that *the cardinality of a linearly independent subset of V cannot exceed that of S.*

Now let us return to the consideration of an arbitrary vector space V over F. A subset S of V is called a *system of generators* of V if every element of V is a linear combination of elements of S. We shall be concerned mostly with vector spaces having a finite system of generators. If V is such a space then the *dimension* of V, denoted by $\dim(V)$, is the smallest natural number m such that there is a system of generators consisting of precisely m elements. Evidently, such a system must be a linearly independent subset, $S = \{s_1, \ldots, s_m\}$ say, of V. With every map f from S to F, we may associate an element

$$\Sigma(f) = f(s_1)s_1 + \cdots + f(s_m)s_m$$

of V, and it is almost evident that Σ is a *linear isomorphism* from the space of all maps from S to F to our space V, i.e., Σ is a group isomorphism compatible with scalar multiplication, in the sense that $\Sigma(\alpha f) = \alpha\Sigma(f)$. Via Σ, our above result concerning the space of maps from S to F implies that the cardinality of a linearly independent subset of V cannot exceed the cardinality m of S. In other words, *if V is a vector space of finite dimension m, then the cardinality of every linearly independent subset of V is at most equal to m.*

A linearly independent system of generators of a vector space V is called a *basis* of V. Using the result just established, it is easy to prove the following facts. *If V is a vector space of finite dimension m, then every basis of V has cardinality m, and every linearly independent subset of V is contained in some basis of V.*

Let U and V be vector spaces over the field F. A *homomorphism of vector spaces*, or simply *linear map*, from U to V is a group homomorphism h from U to V such that $h(\alpha u) = \alpha h(u)$ for every element α of F and every element u of U. Clearly, the value-wise operations make the set of all linear maps from U to V into a vector space over F.

Evidently, every composite of linear maps is a linear map. In particular, in the case where $U = V$, the addition and functional composition of linear maps make the set $\mathrm{End}_F(U)$ of all linear maps from U to U into a ring. Thus, $\mathrm{End}_F(U)$ is simultaneously a ring and a vector space over F. The ring multiplication is compatible with the scalar multiplications, in the sense that, for all elements f and g of $\mathrm{End}_F(U)$ and every element α of F, one has

$$\alpha(f \circ g) = (\alpha f) \circ g = f \circ (\alpha g)$$

One refers to such a simultaneous ring and vector space structure as the structure of an *algebra over F*, or simply an *F-algebra*. Thus, $\mathrm{End}_F(U)$ is an F-algebra in a natural fashion.

2. Let U, V and W be vector spaces over a field F. A *bilinear map* from $U \times V$ to W is a map f with the property that, for every element u of U, the map ${}_uf$ from V to W defined by ${}_uf(v) = f(u, v)$ is a linear map and, for every element v of V, the map f_v from U to W defined by $f_v(u) = f(u, v)$ is also a linear map. In the case where $U = V$ one says that f is *symmetric* if $f(u, v) = f(v, u)$ for all elements u and v of U. In the case where $W = F$, a bilinear map from $U \times U$ to F is also called a *bilinear form on U*.

Now let us consider a vector space V over the field $\mathbf{R}$ of real numbers. We suppose that we are given a symmetric bilinear form f on V which is *positive definite*, in the sense that $f(v, v) > 0$ for every element v of V other than the zero element. As is customary, we abbreviate $f(u, v)$ by $u \cdot v$. When equipped with f, our space V carries a metric geometry, whose points are the elements of V and where the distance between two points u and v is the square root of the non-negative real number $(u - v) \cdot (u - v)$. The fact that this is an appropriate notion of distance comes from the following basic inequality concerning our form f.

For all elements u and v of V, one has

$$(u \cdot v)^2 \leqq (u \cdot u)(v \cdot v)$$

and the equality holds if and only if the pair (u, v) is linearly dependent.

Evidently, it suffices to prove this in the case where neither u nor v is 0. In that case, put $\alpha = (u \cdot u)^{-1/2}$ and $\beta = \varepsilon(v \cdot v)^{-1/2}$, where ε stands for either 1 or -1, selected so as to have $\varepsilon(u \cdot v) \leqq 0$. Now we have

$$0 \leqq (\alpha u + \beta v) \cdot (\alpha u + \beta v) = 2(1 + \alpha\beta u \cdot v)$$

This shows that the absolute value of $u \cdot v$ is no greater than $(\alpha\beta\varepsilon)^{-1}$, which yields $(u \cdot v)^2 \leqq (u \cdot u)(v \cdot v)$. If the equality holds here, we have $\alpha u + \beta v = 0$, showing that the pair (u, v) is linearly dependent. Conversely, if this pair is linearly dependent then one of u or v is a real multiple of the other, which implies that $(u \cdot v)^2 = (u \cdot u)(v \cdot v)$.

Almost inevitably, we regard the 1-dimensional sub vector spaces of V as "lines" of our geometry. Intrinsically, there is nothing to distinguish these subsets from the subsets obtained by adding a fixed point of V to every point of a 1-dimensional sub vector space. Accordingly, we define a *line* of our geometry as a subset of the form $u + \mathbf{R}v$, where u and v are fixed points and $v \neq 0$. Thus, *the lines are the cosets of the* 1-*dimensional sub vector spaces of V*, in the sense of the group structure of V. If p and q are any two distinct points of a line L, then L is the set of points of the form $(1 - \alpha)p + \alpha q$, where α ranges over $\mathbf{R}$. By the *line segment* $[p, q]$ determined by p and q, we mean the set of points $(1 - \alpha)p + \alpha q$ with $0 \leqq \alpha \leqq 1$.

As announced above, we define the distance $D(p, q)$ between the points p and q as the square root of $(p - q) \cdot (p - q)$. The main property of this function D, in addition to having $D(p, q) > 0$ whenever $p \neq q$, is as follows.

For every triple (p, q, r) of elements of V, one has

$$D(p, r) \leqq D(p, q) + D(q, r)$$

and the equality holds if and only if q belongs to the line segment $[p, r]$.

In order to prove this, write u for $p - q$ and v for $q - r$, so that $p - r = u + v$. Generally, if s is any point, write $|s|$ for $D(0, s)$, so that $|s| = (s \cdot s)^{1/2}$. In this notation, we have $D(p, q) = |u|$, $D(q, r) = |v|$ and $D(p, r) = |u + v|$. Now

$$\begin{aligned} |u + v|^2 = (u + v) \cdot (u + v) &= |u|^2 + |v|^2 + 2u \cdot v \\ &\leqq |u|^2 + |v|^2 + 2|u \cdot v| \\ &\leqq |u|^2 + |v|^2 + 2|u||v| \\ &= (|u| + |v|)^2 \end{aligned}$$

This shows that $|u + v| \leqq |u| + |v|$ and that the equality holds if and only if both $u \cdot v = |u \cdot v|$ and $|u \cdot v| = |u||v|$. We know from the above that the second equality holds if and only if $u = \gamma v$ or $v = \gamma u$ for some real number γ. The equality $u \cdot v = |u \cdot v|$ shows that we must then have $\gamma \geqq 0$, except in the case where both u and v are 0, in which case $p = q = r$ and there is nothing to prove. In any case, the equality $D(p, r) = D(p, q) + D(q, r)$ implies that there is a non-negative real number γ such that either

$$q = (1 + \gamma)^{-1}p + \gamma(1 + \gamma)^{-1}r \quad \text{or}$$

$$q = \gamma(1 + \gamma)^{-1}p + (1 + \gamma)^{-1}r$$

In either case, q belongs to $[p, r]$. Finally, one sees directly that if q belongs to $[p, r]$ then one has $D(p, r) = D(p, q) + D(q, r)$.

A structure such as we have just discussed can be attached to any finite-dimensional vector space V over $\mathbf{R}$, as follows. Choose a basis $(v_1, \ldots, v_n)$ of V and define a corresponding bilinear form by the formula

$$\left(\sum_{i=1}^{n} \alpha_i v_i\right) \cdot \left(\sum_{i=1}^{n} \beta_i v_i\right) = \sum_{i=1}^{n} \alpha_i \beta_i$$

Evidently, this form is symmetric and positive definite. Once the basis $(v_1, \ldots, v_n)$ has been chosen, one may identify the elements $\sum_{i=1}^{n} \alpha_i v_i$ of V with the the n-tuples $(\alpha_1, \ldots, \alpha_n)$ of real numbers. Usually, one denotes this vector space of n-tuples by $\mathbf{R}^n$, and one calls this space, equipped with the above bilinear form, *Euclidean n-space*. The form is referred to as the *inner product* of $\mathbf{R}^n$. Now its defining formula reads

$$(\alpha_1, \ldots, \alpha_n) \cdot (\beta_1, \ldots, \beta_n) = \alpha_1\beta_1 + \cdots + \alpha_n\beta_n$$

Let V be again an arbitrary vector space over $\mathbf{R}$, equipped with a symmetric bilinear positive definite form, i.e., an inner product. We wish to discuss the group of bijective distance preserving maps from V to V, i.e., the group of *congruences* of our geometry. It is clear from the definition of distance that every map T_u, where u is a fixed element of V and $T_u(v) = u + v$

for every element v of V, is distance preserving. We refer to T_u as the *translation by u*. It remains to discuss the distance preserving maps that keep 0 fixed.

Let T be such a map, so that $T(0) = 0$ and

$$(T(u) - T(v)) \cdot (T(u) - T(v)) = (u - v) \cdot (u - v)$$

for all elements u and v of V. Expanding this, and then observing that $T(u) \cdot T(u) = u \cdot u$ and $T(v) \cdot T(v) = v \cdot v$, we find $T(u) \cdot T(v) = u \cdot v$. Now we can show that T is a linear map. Let α and β be real numbers. We must show that $T(\alpha u + \beta v) - \alpha T(u) - \beta T(v) = 0$. If we write down the inner product of this element with itself and expand according to bilinearity, we see from the above that the resulting real number remains unchanged when the symbol T is deleted everywhere. But the expression resulting from this deletion is the expanded form of the inner product of $\alpha u + \beta v - \alpha u - \beta v$ with itself, i.e., the resulting expression equals 0. Our conclusion is that T is a linear map.

Thus, *the congruences of V keeping* 0 *fixed are precisely the bijective linear maps T from V to V that preserve the inner product, in the sense that one has* $T(u) \cdot T(v) = u \cdot v$ *for all elements u and v of V*. These maps are called the *orthogonal linear transformations of V*.

The word *orthogonal* here alludes to the relation of orthogonality among elements of V, as well as among lines in V. If u and v are elements of V, one says that u and v are mutually orthogonal if $u \cdot v = 0$. One says that lines L and M are mutually orthogonal if, for all pairs (p, q) of points of L and all pairs (r, s) of points of M, the points $p - q$ and $r - s$ are mutually orthogonal. If v is an arbitrary point of V, then there is exactly one point v_L in L such that $D(v, v_L)$ is the minimum of the set of distances $D(v, w)$ for w in L. If p and q are distinct points of L, this point is characterized by the fact that $p - q$ and $v - v_L$ are mutually orthogonal or, equivalently, that

$$D(v, v_L + \alpha(p - q)) = D(v, v_L - \alpha(p - q))$$

for all real numbers α. If one draws a figure showing these points, one sees that orthogonality in the present sense is in accord with the intuitive geometrical notion of orthogonality or perpendicularity.

Let p be any element of V other than 0. The elements q of V such that $p \cdot q = 0$ evidently constitute a sub vector space V_p of V, which is called the *hyperplane* (through 0) *orthogonal to p*. It is easy to see that every element of V can be written in one and only one way as a sum $\alpha p + q$, where α is a real number and q belongs to V_p. The map from V to V that sends every $\alpha p + q$ to $-\alpha p + q$, i.e., the unique linear map from V to V that leaves the points of V_p fixed and sends p to $-p$, is clearly an *orthogonal* linear map. It is called the *reflection in* V_p. We shall see eventually that, *if V is of finite dimension n, then every orthogonal linear transformation of V is a composite of m reflections, with* $m \leqq n$, *and n is the best possible bound here.*

3. Let us consider the real plane $\mathbf{R}^2$ with its canonical inner product

$$(\alpha_1, \alpha_2) \cdot (\beta_1, \beta_2) = \alpha_1\beta_1 + \alpha_2\beta_2.$$

First, we determine the orthogonal linear transformations of $\mathbf{R}^2$. Let T be such a transformation, put $p = T(1, 0)$ and $q = T(0, 1)$. From the fact that T preserves inner products, we have $p \cdot p = 1 = q \cdot q$ and $p \cdot q = 0$. We write $p = (\sigma, \tau)$, and we note that the inner product relations just written leave us with exactly two possibilities. These are $q = \varepsilon(\tau, -\sigma)$, where ε stands for either 1 or -1.

First, suppose that $\varepsilon = 1$, so that $q = (\tau, -\sigma)$. We observe that each of the two points $(1 + \sigma, \tau)$ and $(\tau, 1 - \sigma)$ is left fixed by T and that at least one of them, r say, is not $(0, 0)$. Now it is easy to verify that T is the reflection in the "hyperplane" $\mathbf{R}r$.

Next, consider the remaining case $\varepsilon = -1$, meaning that $q = (-\tau, \sigma)$. Our transformation T is now given by the formula

$$T(\alpha, \beta) = (\sigma\alpha - \tau\beta, \sigma\beta + \tau\alpha)$$

The remarkable fact about this formula is that it defines a multiplication on $\mathbf{R}^2$, with which $\mathbf{R}^2$ becomes a field containing a copy of the field $\mathbf{R}$ of real numbers, via the map sending each element α of $\mathbf{R}$ to $(\alpha, 0)$. The product of two arbitrary elements (α_1, α_2) and (β_1, β_2) of $\mathbf{R}^2$ is given by the formula

$$(\alpha_1, \alpha_2)(\beta_1, \beta_2) = (\alpha_1\beta_1 - \alpha_2\beta_2, \alpha_1\beta_2 + \alpha_2\beta_1)$$

and our above description of T may now be expressed by saying that T is the multiplication by $T(1, 0) = (\sigma, \tau)$.

With the field structure just described, $\mathbf{R}^2$ is the *field of complex numbers*, which we shall frequently denote by $\mathbf{C}$. For each real number α, one identifies α with $(\alpha, 0)$. Moreover, one usually writes i for $(0, 1)$, so that

$$(\alpha, \beta) = \alpha + \beta i$$

The multiplication of $\mathbf{C}$ is completely determined by the bilinearity of the product with respect to $\mathbf{R}$, together with the fact that $ii = -1$.

There is a ring homomorphism from $\mathbf{C}$ to $\mathbf{C}$, called the *complex conjugation*, which sends each $\alpha + \beta i$ to $\alpha - \beta i$. The elements left fixed by the complex conjugation are precisely the real numbers. Let us indicate the complex conjugation by $*$, so that $(\alpha + \beta i)^* = \alpha - \beta i$. If u is a non-zero complex number then the reciprocal of u may be expressed simply in terms of the complex conjugation and the reciprocal for real numbers. In fact, we have $u^{-1} = (uu^*)^{-1}u^*$. Note that the complex conjugation is its own inverse: $(u^*)^* = u$.

In the conventional diagrammatic presentation of complex numbers as points of the real plane $\mathbf{R}^2$, our above orthogonal transformation T, the multiplication by the complex number p of absolute value (i.e., distance from the origin) 1, is the counter-clockwise rotation around 0 through the angle formed by the rays from 0 toward 1 and from 0 toward p, in this order. At any rate, an orthogonal transformation of this kind is called a (plane) *rotation.* It is not difficult to see from the above that *the reflections are the composites* (in either order) *of the rotations with the complex conjugation, and that every rotation is the composite of a pair of reflections.*

Let u and v be arbitrary complex numbers. We note that the real part of u^*v is precisely the inner product $u \cdot v$, so that we have

$$u^*v = u \cdot v + \delta(u, v)i$$

where $\delta(u, v)$ is a certain real number. Evidently, the function δ so defined is a bilinear form on $\mathbf{R}^2$, and it is *skew symmetric*, in the sense that $\delta(u, v) = -\delta(v, u)$. Moreover, it is easy to see that u^*v is real if and only if the pair (u, v) is linearly dependent. This means that $\delta(u, v)$ is equal to 0 if and only if the pair (u, v) is linearly dependent.

Let T be a rotation, i.e., the multiplication by a complex number p with $p^*p = 1$. Then we have

$$T(u)^*T(v) = (pu)^*pv = u^*p^*pv = u^*v$$

in particular, $\delta(T(u), T(v)) = \delta(u, v)$. We express this property of δ by saying that *δ is invariant with respect to rotations.*

On the other hand, it is seen directly that, for all complex numbers u and v, one has $\delta(u^*, v^*) = -\delta(u, v)$. It follows from this and the invariance with respect to rotations that $\delta(S(u), S(v)) = -\delta(u, v)$ whenever S is a reflection.

If p and q are two distinct points of $\mathbf{R}^2$ then the *ray from p toward q* is the subset of the line, L say, through p and q consisting of the points $p + \rho(q - p)$ with $\rho \geqq 0$. We say that a point r *lies on the left of the ray from p toward q* if r belongs to that side of the line L which appears on the left when one looks along L from p toward q.

The geometrical significance of δ is that *the absolute value of $\delta(u, v)$ is equal to the area of the parallelogram whose vertices are $0, u, v, u + v$. The real number $\delta(u, v)$ is positive if v lies on the left of the ray from 0 toward u, and it is negative if v lies on the right of that ray. Here, $\mathbf{R}^2$ is so oriented that i lies on the left of the ray from 0 toward 1.*

In order to prove this, we first observe that the definitions give

$$(u^*v)^*(u^*v) = (u \cdot v)^2 + \delta(u, v)^2$$

The product on the left equals $v^*uu^*v = (u \cdot u)(v \cdot v)$, so that we have

$$\delta(u, v)^2 = (u \cdot u)(v \cdot v) - (u \cdot v)^2$$

On the other hand, assuming (as we may without loss) that $u \neq 0$, let γu be the point of $\mathbf{R}u$ such that $v - \gamma u$ is orthogonal to u. Then $\gamma = (u \cdot u)^{-1}(v \cdot u)$. The square of the distance from v to the line $\mathbf{R}u$, i.e., to the point γu, is equal to $(v - \gamma u) \cdot (v - \gamma u) = v \cdot v - \gamma u \cdot v$. The product of this and $u \cdot u$ is the square of the area of the parallelogram determined by u and v, and this agrees with the above expression for $\delta(u, v)^2$.

Consider the pair (u, iu). One verifies directly that $\delta(u, iu) = u \cdot u > 0$. Separately treating the eight possible combinations of positivity, negativity or vanishing of the components of u, one checks that iu is on the left of the ray from 0 toward u in every case (by convention, this is true in the case where $u = 1$). Now observe that the line segment $[p, q]$, where p and q are points not on $\mathbf{R}u$, has a point in common with $\mathbf{R}u$ if and only if p and q lie on opposite sides of the ray from 0 toward u. Therefore, our assertion concerning the positivity or negativity of $\delta(u, v)$ will be established as soon as we have shown that $[iu, v]$ meets $\mathbf{R}u$ if and only if $\delta(u, v) < 0$. By definition, this segment consists of the points $\rho iu + (1 - \rho)v$, with $0 \leqq \rho \leqq 1$. We have

$$\delta(u, \rho iu + (1 - \rho)v) = \rho\delta(u, iu) + (1 - \rho)\delta(u, v)$$

We know that this is equal to 0 if and only if $\rho iu + (1 - \rho)v$ belongs to $\mathbf{R}u$. Since $\delta(u, iu) > 0$, the above expression equals 0 for some ρ in our interval if and only if $\delta(u, v) \leqq 0$. This is the required conclusion, because we are not concerned with the case where $\delta(u, v) = 0$, i.e., where v belongs to $\mathbf{R}u$.

Actually, the above function δ is the usual *determinant function*, which is defined, for an arbitrary base field, by the formula

$$\delta((\alpha_1, \alpha_2), (\beta_1, \beta_2)) = \alpha_1\beta_2 - \alpha_2\beta_1$$

This general determinant function is clearly bilinear and skew symmetric, and one has $\delta(p, q) = 0$ if and only if the pair (p, q) is linearly dependent.

4. The determinant function δ on $\mathbf{R}^2 \times \mathbf{R}^2$ is the appropriate tool for calculating areas of closed plane polygons. From the above, it is clear that, if p, q, r are points of $\mathbf{R}^2$, then the absolute value of $\delta(q - p, r - p)$ is equal to twice the area of the triangle whose vertices are p, q, r, and that (assuming that p, q, r are not collinear) $\delta(q - p, r - p)$ is positive or negative according to whether the closed path formed by the line segments $[p, q]$, $[q, r]$, $[r, p]$, in this order, has the counter-clockwise or the clockwise sense.

From the formal properties of δ, we see that

$$\delta(q - p, r - p) = \delta(p, q) + \delta(q, r) + \delta(r, p)$$

Let us consider an arbitrary closed polygon in $\mathbf{R}^2$, given by a vertex sequence $(p_1, \ldots, p_n)$. For notational convenience, we define $p_{n+1} = p_1$. The above result suggests that the sum $\sum_{i=1}^{n} \delta(p_i, p_{i+1})$ has an interpretation in terms of the areas of the regions surrounded by our polygon. We make no special assumptions concerning our closed polygon; it may intersect itself, and the vertices p_i need not be pairwise distinct.

Quite generally, a non-void subset S of Euclidean n-space $\mathbf{R}^n$ is called a *region* if it satisfies the following two conditions.

(1) If p belongs to S then there is a positive real number ε such that every point of $\mathbf{R}^n$ whose distance from p is less than ε belongs to S.

(2) If p and q belong to S then there are points $p = s_1, s_2, \ldots, s_k = q$ such that each line segment $[s_i, s_{i+1}]$ is entirely contained in S.

If our above closed polygon is deleted from $\mathbf{R}^2$ there remains a union of a finite family of pairwise disjoint regions. Let S be one of these regions, and let s be a point of S. For each index i, let R_i denote the ray from s toward p_i. Let α_i be a numerical measure of the angle formed by R_i and R_{i+1}, with α_i positive or negative according to whether p_{i+1} lies on the left or on the right of R_i, the scale being such that the numerical measure of a full counterclockwise turn is 1 (thus, our angular measure is the radian measure divided by 2π). Now $\sum_{i=1}^{n} \alpha_i$ is an integer, which we denote by $I(s)$, and which we call the *winding index* of s with respect to our closed polygon.

Using the above property (2) of the region S, we see that $I(s)$ *is the same for all points s of S*. In fact, as s moves along an unbroken path within S, each α_i, and hence $I(s)$ can change only gradually, without jumps. Since the minimum possible change in $I(s)$ is 1, it follows that $I(s)$ must remain constant as s moves from p to q along a path of line segments as described in (2). We denote this fixed value of the index function I on the region S by $I(S)$, and we call this the *winding index of the region S* with respect to our closed polygon.

Let s be a point not on the polygon, and let t be an arbitrary point of $\mathbf{R}^2$. Let T_i denote the (possibly degenerate) triangle with vertices t, p_i, p_{i+1}. We define an integer $J_i(s, t)$ as follows. If s does not belong to the interior of T_i then $J_i(s, t) = 0$. If s does belong to the interior of T_i then we put $J_i(s, t) = 1$ if p_{i+1} lies on the left of the ray from t toward p_i, and we put $J_i(s, t) = -1$ if p_{i+1} lies on the right of that ray. Thus, if s does not lie on T_i then $J_i(s, t)$ is the winding index of s with respect to T_i. Now suppose that s does not lie on any T_i and consider the sum of the winding indices $J_i(s, t)$ for $i = 1, \ldots, n$. Evidently, the contributions of the rays from t toward the p_i's cancel out in this sum, and what remains is precisely the sum of the contributions of the edges of our closed polygon. Thus, *if s does not lie on any T_i then*

$$\sum_{i=1}^{n} J_i(s, t) = I(s)$$

For every member S of our family of regions and every subset σ of the index set $(1, \ldots, n)$, let S_σ denote the set of points s of S such that s belongs to the interior of T_i if and only if i belongs to σ. Exactly one of our regions is unbounded. We call this S°, and we choose our reference point t from S°. Note that $I(S^\circ) = 0$. Let S be any one of our regions, and let s be a point of S not belonging to the interior of any T_i. Clearly, either s lies on one of the T_i's or else it can be joined to t by a path of line segments not meeting our polygon. In the second case, s belongs to S°. Thus, if $S \neq S^\circ$, then $S_\varnothing$ is contained in the union of the set of rays from t to the p_i's, and the area of S is equal to the sum of the areas of the S_σ's where σ ranges over the non-empty subsets of the index set.

The area of the interior of T_i is equal to the sum of the areas of the S_σ's, where S ranges over our family of regions and σ ranges over the index sets containing i. On the other hand, twice the area of the interior of T_i is equal to the absolute value of $\delta(p_i - t, p_{i+1} - t)$, and if s is a point of the interior of T_i not belonging to our polygon then $J_i(s, t)$ is equal to 1 or -1 according to whether $\delta(p_i - t, p_{i+1} - t)$ is positive or negative.

If S_σ is non-empty, choose a point $p(S, \sigma)$ from S_σ that does not lie on any T_i. Otherwise, for notational convenience, put $p(S, \sigma) = t$. Then our above remarks show that

$$\delta(p_i - t, p_{i+1} - t) = 2 \sum_{S, \sigma} J_i(p(S, \sigma), t)[S_\sigma]$$

where $[S_\sigma]$ stands for the area of S_σ. Summing for i, we derive from this that

$$\sum_{i=1}^{n} \delta(p_i - t, p_{i+1} - t) = 2 \sum_{S, \sigma} I(S)[S_\sigma]$$

For each fixed S, the sum $\sum_\sigma I(S)[S_\sigma]$ is equal to $I(S)[S]$, because the sum of the $[S_\sigma]$'s is $[S]$, except in the case where $S = S^\circ$, in which case $I(S) = 0$. Consequently, our result is

$$\sum_{i=1}^{n} \delta(p_i - t, p_{i+1} - t) = 2 \sum_{S} I(S)[S]$$

where we interpret $I(S^\circ)[S^\circ]$ as 0.

For this final result, the choice of t is immaterial. Indeed, it follows from the formal properties of the determinant function δ that the above sum of values of δ is equal to $\sum_{i=1}^{n} \delta(p_i, p_{i+1})$, for *every* point t. We may state our result as follows.

Let P be the oriented closed polygon determined by the vertex sequence $(p_1, \ldots, p_n)$ in $\mathbf{R}^2$. Define $p_{n+1} = p_1$. Then

$$\sum_{i=1}^{n} \delta(p_i, p_{i+1}) = 2 \sum_{S} I(S)[S]$$

where the sum on the right goes over the finite family of regional components of the complement of P in $\mathbf{R}^2$, and where $I(S)$ is the winding index of S with respect to P, while $[S]$ denotes the area of S.

It is of interest to observe that *winding indices can be determined without the use of any numerical measure for angles*. In order to see this, consider two points p and q not on the polygon whose distances from an edge $[p_i, p_{i+1}]$ are very small compared with their distances from p_i and p_{i+1}. Assume that p lies on the right of this edge, while q lies on the left of it. Then the contribution of this edge to the angle sum at p is nearly one half of a clockwise turn, while its contribution to the angle sum at q is nearly one half of a counterclockwise turn. If, moreover, p and q are close to each other, compared with their distances from all the other edges of the polygon, then it follows that the winding index of q exceeds that of p by exactly 1. This consideration determines the difference of the winding indices of two regions sharing a boundary edge contained in only one polygon edge $[p_i, p_{i+1}]$. The case where a common boundary edge is contained in several polygon edges can evidently be treated by adding the several corresponding differences of the winding indices.

An actual calculation of the winding index of some point p can be made as follows. Construct an arbitrary ray T with p as source. Starting with $I = 0$, add 1 to I for each crossing of T by the polygon from right to left, and subtract 1 from I for each such crossing from left to right. The total is the winding index $I(p)$. The precise meaning of a crossing from right to left, for example, is as follows. Suppose that p_i is on the right of T, and let i' be the first index following i in the cyclic ordering such that $p_{i'}$ is not on T. This counts as a crossing from right to left if $p_{i'}$ is on the left of T, and it does not count as any crossing if $p_{i'}$ is on the right of T.

Exercises

In the following two exercises, for any two distinct points x and y of a vector space, $L(x, y)$ denotes the line containing x and y.

1. Let p, q, r be three non-collinear points in $\mathbf{R}^n$, and let α, β, γ be positive real numbers less than 1. Let u, v, w be the points on the edges of the triangle with vertices p, q, r given by

$$u = \alpha q + (1 - \alpha)r, \qquad v = \beta r + (1 - \beta)p, \qquad w = \gamma p + (1 - \gamma)q$$

Show that the three lines $L(p, u)$, $L(q, v)$, $L(r, w)$ are concurrent if and only if $\alpha\beta\gamma = (1 - \alpha)(1 - \beta)(1 - \gamma)$.

2. Suppose that a, b, c are pairwise linearly independent points in $\mathbf{R}^n$. Let α, β, γ be pairwise distinct real numbers. Show that $L(b, c) \cap L(\beta b, \gamma c) = \{x\}$, $L(c, a) \cap L(\gamma c, \alpha a) = \{y\}$ and $L(a, b) \cap L(\alpha a, \beta b) = \{z\}$, where x, y, z are *collinear* points.

3. A lens changes the direction of light rays passing through it in such a way that all rays issuing from a point sufficiently far in front of the lens continue so as to intersect at a certain corresponding *image point* behind the lens. These image points are determined by the following two conditions.

(1) The direction of any ray passing through the *optical center* of the lens remains unchanged.

(2) The direction of every ray parallel to the *optical axis* of the lens is changed in such a way that the ray passes through a point on the optical axis behind the lens at a certain fixed distance, the *focal length*, from the optical center.

Take the optical center as the origin of $\mathbf{R}^3$, denote the point described in (2) by f, and use the focal length as the unit of length. Now show that the image of every point p with $p \cdot f < -1$ is the point $(1 + p \cdot f)^{-1}p$. Deduce that the image of every line segment whose points satisfy this condition is a line segment and that, in general, the two lines determined by these segments intersect at a point orthogonal to f. What are the exceptions?

4. Suppose that A and B are finite-dimensional sub vector spaces of a vector space V. Let $A + B$ stand for the sub vector space of V consisting of the sums $a + b$, with a in A and b in B. Prove that

$$\dim(A + B) + \dim(A \cap B) = \dim(A) + \dim(B)$$

5. Let V be a finite-dimensional $\mathbf{R}$-space, equipped with an inner product, and let U be a sub vector space of V, with $U \neq V$. Show that there is a point $p \neq 0$ in V such that U is contained in the hyperplane V_p orthogonal to p [choose a basis $(u_1, \ldots, u_m)$ of U, and use **4** above to show that $V_{u_1} \cap \cdots \cap V_{u_m}$ contains a point p as required]. Now let U' denote the sub vector space of V consisting of all points p such that $U \subset V_p$, where we interpret V_0 as V. Prove that $V = U + U'$ and $U \cap U' = \{0\}$. This space U' is called the *orthogonal complement* of U in V.

6. If W is an $\mathbf{R}$-space with an inner product then a basis B of W is called an *orthogonal basis* if $p \cdot q = 0$ for all pairs (p, q) of distinct elements of B. With U and V as in **5** above, show that U has an orthogonal basis, and that every orthogonal basis of U is part of an orthogonal basis of V.

7. Let V be as in **5** above, with $\dim(V) = n > 0$. Suppose that T is an orthogonal linear transformation of V. If $n \leqq 2$, we know already that T is a composite of at most n reflections. Suppose that $n > 2$, and that this result has already been established in the lower cases. Deduce the result for the present n (and thus generally) by the following considerations.

If $T(v)$ is a scalar multiple of v for every point v of V, let $(v_1, \ldots, v_n)$ be an orthogonal basis of V, and obtain the required result for T by noting that, for each i, $T(v_i)$ is one of v_i or $-v_i$.

It remains to deal with the case where there is a point p in V such that the pair $(p, T(p))$ is linearly independent. Put $W = V_p \cap V_{T(p)}$. Show that $\dim(W) = n - 2$, and that $W + \mathbf{R}(p + T(p)) = V_{p-T(p)}$. Let S be the reflection in this hyperplane and show that $S(p) = T(p)$. Now put $U = S^{-1}T$ and show that U maps the hyperplane V_p to itself. Next, apply the inductive hypothesis to the restriction of U to V_p to deduce the required result for T.

Finally, show by induction on n that the map sending each v to $-v$ is not the composite of fewer than n reflections.

8. Let V be a vector space, U a subgroup of V. Show that the factor group V/U can be made into a vector space such that the canonical map from V to V/U becomes a linear map if and only if U is a sub vector space of V. In that case, the vector space V/U is called the *factor space* of V with respect to U. Discuss this more fully, along the line of the discussion of factor groups in Section II.3.

Projects

1. Suppose that p, q, u, v are four distinct points in $\mathbf{R}^2$. Devise computational checks, based on the determinant function and the inner product, for incidence relations involving the line segments $[p, q]$ and $[u, v]$, such as the following: $[p, q] \cap [u, v]$ is empty, or a single point other than p, q, u, v, or u belongs to $[p, q]$, etc. For example, u belongs to $[p, q]$ if and only if $\delta(p - u, q - u) = 0$ and $(p - u) \cdot (q - u) < 0$.

2. Use subroutines obtained in **1** above for sketching a computer program for the determination of winding indices by the method described at the end of Section 4.

3. An electronic plotter accepts computer commands to move the pen one (visually minimal) unit of length up, down, right or left. Make a computer program accomplishing the following task. For each input (h, v), where h and v are integers, the plotter is to draw the visually best approximation to the line segment joining the present pen position to the point situated h units to the right and v units above it. Convince yourself that no recursive program will be acceptable.

CHAPTER V

Multilinear Algebra

1. Let H be an arbitrary set, and let S denote the set whose elements are the finite sequences $(h_1, \ldots, h_n)$ of elements of H, where n ranges over the set of all positive integers. We include in S also the empty sequence ( ), corresponding to the case $n = 0$. We make S into a monoid by defining the composition map simply as juxtaposition, so that the composite of $(h_1, \ldots, h_m)$ and $(k_1, \ldots, k_n)$ is the sequence $(h_1, \ldots, h_m, k_1, \ldots, k_n)$. The empty sequence is the neutral element for this composition.

Now let R be a ring, and let $R[S]$ denote the ring of functions from S to R whose multiplication comes from the monoid composition of S, as explained in Section II.7 (evidently, our monoid S is of finite type). With every element h of H, we may associate the element f_h of $R[S]$, where $f_h((h)) = 1$, while $f_h(s) = 0$ for every other element s of S. Let R_H be the subring of $R[S]$ that is generated by these elements f_h and the identity element of $R[S]$, so that R_H is the smallest subring of $R[S]$ containing all the functions f_h. It is easy to see that the elements of R_H are precisely those functions whose values are 0 outside some finite subset of S, depending on the function, and that these functions are the sums of products of functions f_h and elements of R. In fact, $f_{h_1} \cdots f_{h_n}$ is the *characteristic function* of the sequence $(h_1, \ldots, h_n)$, i.e., it takes the value 1 at this sequence and the value 0 at every other element of S.

From now on, we suppose that our ring R is commutative. In Section IV.1, we introduced the notions of an R-linear map and an R-algebra, in the case where R is a field. Evidently, these notions extend without change to the case where R is an arbitrary commutative ring. If A and B are R-algebras, then a *homomorphism of R-algebras* from A to B is a ring homomorphism that is also an R-linear map.

In these terms, the significance of our above construction of R_H resides in the following fact. *Let ρ denote the map from H to R_H given by $\rho(h) = f_h$. For* *every map α from H to an R-algebra A, there is one and only one homomorphism α^* of R-algebras from R_H to A such that $\alpha^* \circ \rho = \alpha$.*

Essentially, the proof of this statement is dictated by the statement itself. It involves no difficulties once one realizes that, if f is any element of R_H, then $\alpha^*(f)$ is the sum in A of the finite set of terms $f((h_1, \ldots, h_n))\alpha(h_1) \cdots \alpha(h_n)$ in which $f((h_1, \ldots, h_n))$ is different from 0. For $n = 0$, the summand is $f((\ ))1_A$, where 1_A stands for the identity element of the ring A. We refer to this feature of (R_H, ρ) as the *universal mapping property*.

We adapt the above construction to multilinear algebra, as follows. Suppose that U is a vector space over a field F, and consider the F-algebra F_U. If α is a linear map from U to an F-algebra A, then the kernel of the F-algebra homomorphism α^* from F_U to A contains every element of the form $\rho(\sigma u) - \sigma\rho(u)$, where σ is an element of F, and also every element of the form $\rho(u_1 + u_2) - \rho(u_1) - \rho(u_2)$. Let I_U denote the ideal of F_U whose elements are the sums of products of elements of F_U by elements of these two types. Then I_U is contained in the kernel of every α^* as above. Accordingly, we introduce the factor F-algebra F_U/I_U, which we denote by $T(U)$, and which we call the *tensor F-algebra built on U*.

Let τ stand for the map from U to $T(U)$ given by $\tau(u) = \rho(u) + I_U$. It follows from the definition of I_U that τ is a *linear* map. From our above discussion, it is now clear that $(T(U), \tau)$ has the following universal mapping property. *For every linear map α from U to an F-algebra A, there is one and only one homomorphism $T(\alpha)$ of F-algebras from $T(U)$ to A such that $T(\alpha) \circ \tau = \alpha$.*

Now let B be a basis of U. It is clear from the above definition that every element of $T(U)$ is an F-linear combination of the identity element of $T(U)$ and the products $\tau(b_1) \cdots \tau(b_n)$ corresponding to the finite sequences $(b_1, \ldots, b_n)$ of elements of B. We claim that *these products, together with the identity element of $T(U)$, actually constitute a basis of $T(U)$ as a vector space over F*, i.e., that these elements are pairwise distinct and constitute a linearly independent system of F-space generators of $T(U)$. In order to prove this, we construct the F-algebra F_B, and we let ρ now denote the associated map from B to F_B. Let α be the linear map from U to F_B that sends every element b of B to $\rho(b)$. Then the corresponding map $T(\alpha)$ from $T(U)$ to F_B sends the identity element of $T(U)$ to that of F_B and every product $\tau(b_1) \cdots \tau(b_n)$ to the product $\rho(b_1) \cdots \rho(b_n)$ in F_B. It is evident from the definition of F_B that these products are pairwise distinct from each other and from the identity element of F_B, and that they constitute a linearly independent set. Since $T(\alpha)$ is a linear map, this implies that the same is true for the antecedents of these elements in $T(U)$.

Whenever the choice of a particular basis B of U does not cloud the issue unduly, the use of the above corresponding basis of $T(U)$ facilitates computational control in this context to a considerable extent. In this way, $T(U)$ appears as the *F-algebra of polynomials in the elements of B as free, non-commuting symbols.*

There are several important factor algebras of the tensor algebra whose construction is motivated by various specializations of linear maps from U to F-algebras. For our present purposes, the most important factor algebra of $T(U)$ is the *exterior F-algebra built on U, which we denote by $E(U)$*. This is defined as $T(U)/J_U$, where J_U is the ideal generated, as such, by the elements $\tau(u)^2$. We define the linear map ε from U to $E(U)$ by $\varepsilon(u) = \tau(u) + J_U$, and we note that $(E(U), \varepsilon)$ has the following universal mapping property. *For every linear map α from U to an F-algebra A having the property that $\alpha(u)^2 = 0$ for every element u of U, there is one and only homomorphism $E(\alpha)$ of F-algebras from $E(U)$ to A such that $E(\alpha) \circ \varepsilon = \alpha$.*

Note that $\varepsilon(u)^2 = 0$ for every element u of U. Using this with $u = u_1 + u_2$, we see that this implies that $\varepsilon(u_1)\varepsilon(u_2) = -\varepsilon(u_2)\varepsilon(u_1)$ for all elements u_1 and u_2 of U. It follows from this that if B is a *totally ordered* basis of U then every element of $E(U)$ is a linear combination of the identity element of $E(U)$ and the products $\varepsilon(b_1) \cdots \varepsilon(b_n)$ *with $b_1 < \cdots < b_n$ in the given ordering of B*. We shall show that *these elements are pairwise distinct and constitute a basis of $E(U)$*.

In order to do this, we must look more closely at $T(U)$ and J_U. For every non-negative integer d, let $T^d(U)$ denote the sub vector space of $T(U)$ consisting of the F-linear combinations of products of d factors from $\tau(U)$. We call $T^d(U)$ the *homogeneous component of degree d* of $T(U)$, agreeing that $T^0(U)$ consists of the F-multiples of the identity element of $T(U)$. Evidently, every element t of $T(U)$ determines an element t_d of each $T^d(U)$ such that t is the sum of the finite set of those t_d's which are different from 0. We refer to t_d as the component of degree d of t, and we call the elements of $T^d(U)$ *homogeneous elements of degree d*. Finally, note that the product set $T^d(U)T^e(U)$ is contained in $T^{d+e}(U)$. We refer to these features of $T(U)$ by saying that $T(U)$ is a *graded* F-algebra.

Next, we note that J_U is a *homogeneous* ideal, in the sense that, if t is an element of J_U, then all its components t_d belong to J_U, so that J_U is the sum of its intersections with the $T^d(U)$'s. Consequently, $E(U)$ inherits a grading from $T(U)$ in the evident way: $E^d(U)$ consists of the cosets $t + J_U$ with t in $T^d(U)$.

Now let us return to our ordered basis B of U. Let η be any map from B to F. We define a corresponding linear map η° from $T(U)$ to $T(U)$ as follows. As a linear map, η° will be determined by its effect on the elements of our basis of $T(U)$ corresponding to B. We require that η° map the identity element of $T(U)$ to 0, so that $\eta^\circ(T^0(U)) = \{0\}$. For the other basis elements $\tau(b_1) \cdots \tau(b_n)$, we define

$$\eta^\circ(\tau(b_1) \cdots \tau(b_n)) = \sum_{i=1}^{n} (-1)^{i-1}\eta(b_i)[\tau(b_1) \cdots \tau(b_n)]_i$$

where $[\tau(b_1) \cdots \tau(b_n)]_i$ stands for the product remaining when the factor $\tau(b_i)$ is deleted from $\tau(b_1) \cdots \tau(b_n)$. For $n = 1$, we agree that the remaining empty product stands for the identity element of $T(U)$. Clearly, η° has the following properties. For every positive index d, we have $\eta^\circ(T^d(U)) \subset T^{d-1}(U)$, while $\eta^\circ(T^0(U)) = \{0\}$. If x belongs to $T^d(U)$ and y is an arbitrary element of $T(U)$ then

$$\eta^\circ(xy) = \eta^\circ(x)y + (-1)^d x\eta^\circ(y)$$

One expresses these formal properties by saying that η° is a *homogeneous F-algebra derivation of degree* -1.

Now observe that J_U is the ideal generated by the elements of the form $\tau(b)^2$ and the elements of the form $\tau(b_1)\tau(b_2) + \tau(b_2)\tau(b_1)$, where b, b_1 and b_2 range over B. Evidently, these elements belong to the kernel of η°. From the formula for $\eta^\circ(xy)$ given above, we see now that $\eta^\circ(J_U) \subset J_U$.

Our assertion concerning a basis of $E(U)$ corresponding to B will be proved as soon as we have shown that a linear combination of the identity element of $T(U)$ and products $\tau(b_1)\cdots\tau(b_n)$ with $b_1 < \cdots < b_n$ belongs to J_U only if each coefficient is 0. Since $T(U)$ is graded and J_U is homogeneous, it suffices to deal with the case where this linear combination is homogeneous, i.e., where the number n of factors is the same in each term. Let $b_1, \ldots, b_n$ be as just above. For each i from $(1, \ldots, n)$, let η_i be the map from B to F whose value at b_i is 1 and all whose other values are 0. Write β_i for $(\eta_i)^\circ$. One verifies directly that the composite map $\beta_n \circ \cdots \circ \beta_1$ sends $\tau(b_1)\cdots\tau(b_n)$ to the identity element of $T(U)$, while it annihilates every other such product $\tau(b'_1)\cdots\tau(b'_n)$ with $b'_1 < \cdots < b'_n$. If our linear combination belongs to J_U, then its image under our composite map belongs to the homogeneous component of degree 0 of J_U, which is $\{0\}$. Therefore, the coefficient of $\tau(b_1)\cdots\tau(b_n)$ in the linear combination must be 0. The only case not covered by this argument is the case $n = 0$, in which case there is nothing to prove, in view of the fact that the component of degree 0 of J_U is $\{0\}$.

2. The most significant feature of the exterior algebra is that it embodies automatic control of linear dependence. Let U be a vector space over a field F. It is clear from the above that the map ε from U to $E(U)$ is injective. Accordingly, we use this map to identify U with the corresponding sub vector space $E^1(U)$ of $E(U)$. Now our notation is simplified by writing u instead of $\varepsilon(u)$ for every element u of U. In this notation, we have the following linear dependence criterion. *A sequence $(u_1, \ldots, u_k)$ of elements of U is linearly dependent if and only if the product $u_1 \cdots u_k$ in $E(U)$ is equal to* 0.

In order to see this, let us first suppose that the sequence is linearly dependent. Then, for at least one index i, the element u_i is a linear combination of the u_j's with $j \neq i$. If we substitute this linear combination for u_i in the product $u_1 \cdots u_k$, we obtain a linear combination of products in each of which one of the u_j's occurs twice as a factor. Therefore, the product is equal to 0. On the other hand, if the sequence $(u_1, \ldots, u_k)$ is not linearly dependent then, as written, it is an initial segment of an ordered basis of U. Consequently, the product $u_1 \cdots u_k$ is an element of the basis of $E(U)$ corresponding to such an ordered basis of U. In particular, it is not equal to 0.

This criterion is decisive for the use of the exterior algebra in the theory of systems of linear equations, as follows. Let us consider a system of n equations in m unknowns $\eta_1, \ldots, \eta_m$ (to be found in the base field F) with coefficients in F

$$\sum_{j=1}^{m} \alpha_{ij}\eta_j = \gamma_i \qquad (i = 1, \ldots, n)$$

Let F^n denote the F-space of n-tuples of elements of F, which is now to take the place of the vector space U above. For each j from $(1, \ldots, m)$, let $a_j = (\alpha_{1j}, \ldots, \alpha_{nj})$, and put $c = (\gamma_1, \ldots, \gamma_n)$. The given system of equations is equivalent to the single condition $\sum_{j=1}^{m} \eta_j a_j = c$ in F^n.

Working in $E(F^n)$, we begin by determining the largest index r (which cannot exceed n) for which at least one of the products $a_{j_1} \cdots a_{j_r}$ is different from 0. Then, let us relabel the unknowns η_j, and correspondingly the α_{ij}'s, so that $a_1 \cdots a_r \neq 0$. Since every product of more than r of the a_j's is 0, we know from the above linear dependence criterion that each a_j is a linear combination of $a_1, \ldots, a_r$. Therefore, a necessary and sufficient condition for the existence of a solution $(\eta_1, \ldots, \eta_m)$ is that c be a linear combination of $a_1, \ldots, a_r$. By the above criterion, this condition is simply $a_1 \cdots a_r c = 0$.

Now let us assume that this condition is satisfied, and let us find the set of all solutions. Our condition for the η_j's may be written in the form

$$\sum_{j=1}^{r} \eta_j a_j = c - \sum_{j=r+1}^{m} \eta_j a_j$$

Fix $\eta_{r+1}, \ldots, \eta_m$ arbitrarily in F, and abbreviate the expression on the right side above by b, so that our condition is $\sum_{j=1}^{r} \eta_j a_j = b$. Multiplying from the left by $a_1 \cdots a_{k-1}$ and from the right by $a_{k+1} \cdots a_r$, where k is an index from $(1, \ldots, r)$, we obtain the condition

$$\eta_k a_1 \cdots a_r = a_1 \cdots a_{k-1} b a_{k+1} \cdots a_r$$

Like every product of r elements a_j or c, the product on the right here is a scalar multiple $\beta_k a_1 \cdots a_r$, and our condition means that we must have $\eta_k = \beta_k$ for each k.

Conversely, if $\eta_k = \beta_k$ for each index k from $(1, \ldots, r)$, then each product

$$a_1 \cdots a_{k-1}\left(b - \sum_{j=1}^{r} \eta_j a_j\right) a_{k+1} \cdots a_r$$

is equal to 0. In any case, the middle factor in parentheses is a certain linear combination of $a_1, \ldots, a_r$, say $\sum_{j=1}^{r} \delta_j a_j$. From the vanishing of the products, we deduce that each coefficient δ_j here must be equal to 0, because the product written above is equal to $\delta_k a_1 \cdots a_r$. Therefore, we have $b - \sum_{j=1}^{r} \eta_j a_j = 0$, which means that our original condition is indeed satisfied.

Thus, the complete set of solutions is the family of elements

$$(\beta_1, \ldots, \beta_r, \eta_{r+1}, \ldots, \eta_m) \text{ of } F^m,$$

where the entries $\eta_{r+1}, \ldots, \eta_m$ range independently over F, and where the β_k's are determined from each choice of $\eta_{r+1}, \ldots, \eta_m$ by the equations

$$a_1 \cdots a_{k-1}\left(c - \sum_{j=r+1}^{m} \eta_j a_j\right)a_{k+1} \cdots a_r = \beta_k a_1 \cdots a_r$$

In particular, this shows that the solutions constitute a coset of a sub vector space of F^m of dimension $m - r$.

The above considerations are closely related to the study of a single linear map, t say, from a finite-dimensional F-space U to itself. Regarding t as a linear map from U to the F-algebra $E(U)$, we conclude from the universal mapping property of $E(U)$ that there is one and only one homomorphism $E(t)$ of F-algebras from $E(U)$ to itself whose restriction to U coincides with t. Clearly, $E(t)$ maps each $E^m(U)$ into itself.

In particular, consider the restriction of $E(t)$ to $E^d(U)$, where d is the dimension of U. We know from our above results concerning bases of $E(U)$ that $E^d(U)$ is of dimension 1. In fact, if $(u_1, \ldots, u_d)$ is a basis of U then $E^d(U)$ consists simply of the scalar multiples of the product $u_1 \cdots u_d$. Since the restriction of $E(t)$ to $E^d(U)$ is a linear endomorphism of $E^d(U)$, it is therefore the scalar multiplication by a certain element of F. We denote this element of F by $\delta(t)$, and we call it the *determinant* of t. This defines an F-valued function δ on $\mathrm{End}_F(U)$, which is called the *determinant function.* Clearly, if t_1 and t_2 are linear endomorphisms of U then $E(t_1 \circ t_2) = E(t_1) \circ E(t_2)$. Consequently, δ is a multiplicative function, in the sense that $\delta(t_1 \circ t_2) = \delta(t_1)\delta(t_2)$.

Since $E(t)$ is a homomorphism of algebras, we have

$$E(t)(u_1 \cdots u_d) = t(u_1) \cdots t(u_d)$$

Evidently, t is injective if and only if the sequence $(t(u_1), \ldots, t(u_d))$ is linearly independent. Therefore, *the determinant of a linear endomorphism t is equal to 0 if and only if t is not injective.*

The most important question in the study of a linear endomorphism t concerns the existence of *characteristic vectors* for t, by which one means non-zero elements u of U such that $t(u)$ is a scalar multiple of u. The existence of such an element u means that, if i is the identity map on U, there is some element γ in F such that the kernel of $\gamma i - t$ is not $\{0\}$, i.e., such that $\gamma i - t$ is not injective. Every non-zero element u of the kernel of $\gamma i - t$ is a characteristic vector of t, with $t(u) = \gamma u$. The element γ is then called a characteristic value for t, and u is said to *belong to* the characteristic value γ.

From the above, we see that γ is a characteristic value for t if and only if $\delta(\gamma i - t) = 0$. By choosing a basis $(u_1, \ldots, u_d)$ of U and writing each $t(u_i)$ as a linear combination of u_j's, one sees that $\delta(\gamma i - t)$, as a function of γ, is a polynomial of degree d with coefficients in F and highest coefficient 1. This polynomial is called the *characteristic polynomial* of t. Its roots, in any field containing F, are called characteristic values for t. The characteristic vectors for t in U belong to those characteristic values which lie in F.

3. Let U be a vector space over a field F. The linear maps from U to F constitute a vector space over F, with the value-wise operations. This is called the space *dual* to U, and we denote it by U'. From the analysis we made in Section 1 of the ideal J_U of $T(U)$ defining $E(U)$, we know that every element η of U' determines a homogeneous F-algebra derivation D_η of degree -1 on $E(U)$ such that D_η coincides with η on U, when we identify $E^1(U)$ with U and $E^0(U)$ with F in the usual way. Evidently, the map sending each η to D_η is a linear map from U' to $\text{End}_F(E(U))$. It follows from the effect of D_η on products that $D_\eta^2(xy) = D_\eta^2(x)y + xD_\eta^2(y)$ for all elements x and y of $E(U)$. Since D_η^2 annihilates $E^0(U) + E^1(U)$, it follows that $D_\eta^2 = 0$. Now we conclude from the universal mapping property of $E(U')$ that there is one and only one homomorphism π of F-algebras from $E(U')$ to $\text{End}_F(E(U))$ such that $\pi(\eta) = D_\eta$ for every element η of U'. If γ is an element of $E^m(U')$ then $\pi(\gamma)$ maps each $E^n(U)$ into $E^{n-m}(U)$. In particular, the restriction of $\pi(\gamma)$ to $E^m(U)$ may be regarded as an element of $E^m(U)'$. In this way, π yields a linear map π_m from $E^m(U')$ to $E^m(U)'$.

We claim that each of the linear maps π_m is injective. Evidently, it suffices to deal with the case where $m > 0$. Let γ be an element of $E^m(U')$. From Section IV.1, we see that there are elements $\eta_1, \ldots, \eta_n$ of U' such that γ can be written as a linear combination of the products $\eta_{i_1} \cdots \eta_{i_m}$, with $i_1 < \cdots < i_m$, and to which there correspond elements $u_1, \ldots, u_n$ in U such that, for each index i, we have $\eta_i(u_i) = 1$, while $\eta_i(u_j) = 0$ whenever $j \neq i$. Now one sees readily from the definition of π_m that the value of $\pi_m(\gamma)$ at $u_{i_m} \cdots u_{i_1}$ is the coefficient of $\eta_{i_1} \cdots \eta_{i_m}$ in γ. Clearly, this implies that π_m is injective.

Now let us suppose that U is of finite dimension d. Then U' is of dimension d also. More precisely, let $(u_1, \ldots, u_d)$ be a basis of U. For each index i, let η_i be the element of U' whose value at u_i is 1, while its value at every u_j other than u_i is 0. Then $(\eta_1, \ldots, \eta_d)$ is a basis of U'. It is called the basis *dual* to the basis $(u_1, \ldots, u_d)$. Now $E^m(U')$ and $E^m(U)'$ are of the same dimension $\binom{d}{m}$ (which is to be interpreted as 0 when $m > d$). Since π_m is injective, it follows that *π_m is actually a linear isomorphism whenever U is finite-dimensional.*

Let t be a linear map from U to some vector space V. In the natural way, t gives rise to a linear map t' from V' to U', where $t'(\eta) = \eta \circ t$ for every element η of V'. One calls t' the *transpose* (or also the dual) of t. With regard to the exterior algebras, one verifies directly that $E(t) \circ D_{t'(\eta)} = D_\eta \circ E(t)$, and hence that, for every γ in $E(V')$,

$$E(t) \circ \pi(E(t')(\gamma)) = \pi(\gamma) \circ E(t)$$

In particular, consider the case where U is of finite dimension d and $V = U$. Let γ range over $E^d(U')$. Using that $E(t')(\gamma) = \delta(t')\gamma$, we see from our last result above that $\delta(t')\pi(\gamma) = \pi(\gamma) \circ E(t)$. Via evaluation on $E^d(U)$, this yields $\delta(t') = \delta(t)$. Thus, *for every linear endomorphism t of a finite-dimensional vector space, the determinant of t is equal to the determinant of the transpose of t.*

Let us fix a basis $(u_1, \ldots, u_d)$ of U and the dual basis $(\eta_1, \ldots, \eta_d)$ of U'. Write $t(u_i) = \sum_{j=1}^{d} \gamma_{ji}(t)u_j$, with each $\gamma_{ji}(t)$ in F. Then the square array of the field elements $\gamma_{ji}(t)$, where the first subscript is used as the row index and the second subscript as the column index, is called the *matrix* of the linear endomorphism t with respect to the basis $(u_1, \ldots, u_d)$ of U. By the determinant of such a matrix, one means the determinant of the corresponding linear endomorphism. Evidently, we have

$$\gamma_{ji}(t) = \eta_j(t(u_i)) = t'(\eta_j)(u_i) = \gamma_{ij}(t')$$

where the $\gamma_{ij}(t')$'s are the entries of the matrix of t' with respect to the basis $(\eta_1, \ldots, \eta_d)$ of U'. Thus, *with respect to any given basis of U and the dual basis of U', the matrix of t' is the transpose of the matrix of t*, in the sense that the rows of the matrix of t' are the columns of the matrix of t.

4. We wish to discuss the sub vector spaces of a finite-dimensional vector space over a field F. In order to facilitate references to Section 3, we take our containing vector space to be U', where U is a vector space of dimension d over F. Let m be an index with $0 < m \leqq d$. If S is an m-dimensional sub vector space of U' then the products $\sigma_1 \cdots \sigma_m$ in $E(U')$ with each factor σ_i in S constitute a 1-dimensional sub vector space $[S]$ of $E^m(U')$. In fact, if $(\sigma_1, \ldots, \sigma_m)$ is a basis of S then $[S] = F\sigma_1 \cdots \sigma_m$. Let us say that a non-zero element of $E(U')$ is *simple* if it can be written as a product of elements of U'. The 1-dimensional space of all scalar multiples of a simple element of $E^m(U')$ will be called a *simple subspace.* Then it is clear that the map sending each S as above to $[S]$ is a bijective map from the set of m-dimensional subspaces of U' to the set of simple subspaces of $E^m(U')$.

For this reason, it is of interest to obtain criteria enabling one to decide whether or not a given non-zero element p of $E^m(U')$ is simple. One such criterion involves the dimension of the annihilator of p in U'. Thus, let A_p denote the sub vector space of U' consisting of the elements α such that $\alpha p = 0$. We shall prove that $\dim(A_p) \leqq m$ *for every non-zero element p of $E^m(U')$, and* that p *is simple if and only if* $\dim(A_p) = m$.

Let $(\alpha_1, \ldots, \alpha_k)$ be a basis of A_p, and complete this to a basis $(\alpha_1, \ldots, \alpha_d)$ of U'. Now write

$$p = \sum_{i_1 < \cdots < i_m} \gamma_{i_1 \cdots i_m} \alpha_{i_1} \cdots \alpha_{i_m}$$

with each $\gamma_{i_1 \cdots i_m}$ in F. Let i be an index $\leqq k$. Since $\alpha_i p = 0$, we must have $\gamma_{i_1 \cdots i_m} = 0$ unless one of the subscripts i_r is i. If $\gamma_{i_1 \cdots i_m} \neq 0$, then each index $i \leqq k$ must therefore be one of these subscripts i_j. Since $p \neq 0$, it follows that $k \leqq m$. In the case where $k = m$, our argument clearly gives the conclusion that $p = \gamma_{1 \ldots m} \alpha_1 \cdots \alpha_m$, showing that p is simple.

Conversely, if p is simple, we have $p = \eta_1 \cdots \eta_m$, with each η_i in U'. Since $p \neq 0$, the sequence $(\eta_1, \ldots, \eta_m)$ is linearly independent. Evidently, $\eta_i p = 0$ for each i. Thus, A_p contains each η_i, so that $\dim(A_p) \geqq m$.

We can use this criterion in order to show that *every non-zero element of* $E^{d-1}(U')$ *is simple* (assuming that $d > 1$). Using a basis $(\eta_1, \ldots, \eta_d)$ of U', let p_i stand for the product remaining when η_i is deleted from $\eta_1 \cdots \eta_d$. Write the given element p of $E^{d-1}(U')$ in the form $\sum_{i=1}^d \gamma_i p_i$. Then the condition that an element $\sum_{i=1}^d \rho_i \eta_i$ of U' be in A_p reduces to the single condition

$$\sum_{i=1}^{d} (-1)^{i-1} \gamma_i \rho_i = 0$$

for the coefficients ρ_i. Evidently, this implies that the space of solutions $(\rho_1, \ldots, \rho_d)$ in F^d is of dimension no less than $d - 1$, i.e., $\dim(A_p) \geqq d - 1$. Now our above criterion gives the conclusion that p is simple.

Note that, *in the cases* $m = 1$ *and* $m = d$, *it is obvious that every non-zero element of* $E^m(U')$ *is simple.*

Using the map π from $E(U')$ to $\mathrm{End}_F(E(U))$ discussed in Section 3, we obtain the following computational characterization of the simple elements of $E^m(U')$. *Let g be a generator of the 1-dimensional space $E^d(U)$. A non-zero element p of $E^m(U')$ is simple if and only if*

$$\pi(p)(g)\pi(px)(g) = 0$$

for every element x of $E^{d-m-1}(U')$ (we interpret $E^{-1}(U')$ as $\{0\}$).

First, suppose that p is simple, so that $p = \eta_1 \cdots \eta_m$, where $(\eta_1, \ldots, \eta_m)$ is a linearly independent sequence of elements of U'. We complete this sequence to a basis $(\eta_1, \ldots, \eta_d)$ of U', which is, of course, the dual of some basis $(u_1, \ldots, u_d)$ of U. Without loss, we take $g = u_1 \cdots u_d$. If $x = \eta_{i_1} \cdots \eta_{i_{d-m-1}}$, with $i_1 < \cdots < i_{d-m-1}$, then $px = 0$ unless each i_j is greater than m. If each i_j is greater than m then there is exactly one index k that is greater than m and different from each i_j in our expression for x, and we have $\pi(px)(g) = \pm u_k$. On the other hand, $\pi(p)(g) = \pm u_{m+1} \cdots u_d$. Therefore, we have

$$\pi(p)(g)\pi(px)(g) = 0$$

in every case, and hence for every element x of $E^{d-m-1}(U')$.

Now suppose that p is a non-zero element of $E^m(U')$ satisfying the above vanishing conditions, and write

$$p = \sum_{i_1 < \cdots < i_m} \gamma_{i_1 \cdots i_m} \eta_{i_1} \cdots \eta_{i_m}$$

where $(\eta_1, \ldots, \eta_d)$ is a basis of U' and $\gamma_{1 \cdots m} = 1$. For each index j with $m < j \leqq d$, let x_j be the product that remains when η_j is deleted from $\eta_{m+1} \cdots \eta_d$ (we assume that $m < d$, noting that there is nothing more to prove when $m = d$). We have $px_j\eta_j = \pm\eta_1 \cdots \eta_d$, while $px_j\eta_i = 0$ for every index i that is greater than m and different from j. From this, we see that the sequence $(px_{m+1}, \ldots, px_d)$ is linearly independent. The same is therefore true for the sequence $(\pi(px_{m+1}), \ldots, \pi(px_d))$. It is easy to see that the evaluation at g is an injective map from $\pi(E^{d-1}(U'))$ to U. Consequently, the sequence $(\pi(px_{m+1})(g), \ldots, \pi(px_d)(g))$ in U is linearly independent. For each $j > m$, put $u_j = \pi(px_j)(g)$, and choose $u_1, \ldots, u_m$ from U such that $(u_1, \ldots, u_d)$ is a basis of U. From the assumption on p, we have $\pi(p)(g)u_j = 0$ for every $j > m$. This implies that $\pi(p)(g)$ is a scalar multiple of $u_{m+1} \cdots u_d$. If $(\mu_1, \ldots, \mu_d)$ is the basis of U' dual to the basis $(u_1, \ldots, u_d)$ of U, our last conclusion shows that p is a scalar multiple of $\mu_1 \cdots \mu_m$, so that p is simple.

If p is written as $\sum_{i_1 < \cdots < i_m} \gamma_{i_1 \cdots i_m} \eta_{i_1} \cdots \eta_{i_m}$ where $(\eta_1, \ldots, \eta_d)$ is basis of U', then our criterion says that p is simple if and only if the coefficients $\gamma_{i_1 \cdots i_m}$ constitute a common zero of a certain family of homogeneous polynomials of degree 2 in $\binom{d}{m}$ variables, the number of polynomials in the family being the dimension $\binom{d}{d-m-1} = \binom{d}{m-1}$ of $E^{d-m-1}(U')$.

Exercises

1. Let $T(U)$ be the tensor algebra built on a vector space U over a field F. Let K_U denote the ideal of $T(U)$ that is generated by the elements of the form $\tau(u_1)\tau(u_2) - \tau(u_2)\tau(u_1)$, where u_1 and u_2 range over U and τ is the canonical linear map from U to $T(U)$. The factor algebra $T(U)/K_U$ is called the *symmetric F-algebra built on U*. Let σ stand for the linear map from U to $S(U)$ obtained from τ in the evident way. Formulate and validate a universal mapping property of $(S(U), \sigma)$ analogous to that given in Section 1 for $(E(U), \varepsilon)$. Let B be a basis of U. Call two finite sequences of elements of B equivalent if they differ only in the ordering of the terms. Now show that the set of products $\sigma(b_1) \cdots \sigma(b_n)$ corresponding to the sequences $(b_1, \ldots, b_n)$ together with the identity element of $S(U)$ (corresponding to the empty sequence), is a basis of $S(U)$ each of whose elements corresponds to one and only one equivalence class of sequences in B.

2. Let V be a finite-dimensional $\mathbf{R}$-space with an inner product, and let $(u_1, \ldots, u_m)$ be a finite sequence of elements of V. For each index j, put $v_j = \sum_{i=1}^{m} (u_j \cdot u_i)u_i$. Prove that, in the exterior algebra $E(V)$, one has

$$v_1 \cdots v_m = c(u_1, \ldots, u_m)u_1 \cdots u_m$$

where the real number $c(u_1, \ldots, u_m)$ is the square of the volume of the parallelepiped in V whose vertices are the 2^m sums (without repetitions of summands) formed from $(u_1, \ldots, u_m)$, including the empty sum 0.

Note that there is nothing to prove unless the sequence $(u_1, \ldots, u_m)$ is linearly independent. In that case, show that there is a linear endomorphism T of V such that $T(u_j) = v_j$ for each index j, and that

$$v_1 \cdots v_m = E(T)(u_1 \cdots u_m)$$

Next, show that one may write $u_m = z + h$, where z is a linear combination of the u_i's with $i < m$, while $h \cdot u_i = 0$ whenever $i < m$. Now show that $u_1 \cdots u_m = u_1 \cdots u_{m-1}h$, and deduce that

$$E(T)(u_1 \cdots u_m) = T(u_1) \cdots T(u_{m-1})(h \cdot u_m)u_m$$

Verify that this last product in $E(V)$ is equal to $(h \cdot h)w_1 \cdots w_{m-1}u_m$, where $w_j = \sum_{i=1}^{m-1} (u_j \cdot u_i)u_i$. Finally, use this result for establishing the required result by induction on m.

3. Consider the situation of Exercise 2 in the case where $V = \mathbf{R}^m$. Show that then $c(u_1, \ldots, u_m)$ is equal to the square of the determinant of the matrix whose columns are the m-tuples $u_1, \ldots, u_m$ of real numbers, and conclude that the absolute value of the determinant of the linear endomorphism sending the canonical basis of $\mathbf{R}^m$ to $(u_1, \ldots, u_m)$ is equal to the volume of the parallelepiped determined by $(u_1, \ldots, u_m)$.

You will need the formula for the matrix of the composite of two linear endomorphisms of $\mathbf{R}^m$ in terms of the matrices of the two given endomorphisms, the multiplicative property of the determinant function and the fact that the determinant of a matrix is equal to the determinant of its transpose.

4. Let U be a finite-dimensional $\mathbf{R}$-space with an inner product. Note that the inner product yields an isomorphism, μ say, from U to U', where $\mu(x)(y) = x \cdot y$ for all elements x and y of U. Now μ yields the isomorphism $E(\mu)$ of $\mathbf{R}$-algebras from $E(U)$ to $E(U')$. Composing this with the homomorphism π of Section 3, one obtains a homomorphism $\gamma = \pi \circ E(\mu)$ of $\mathbf{R}$-algebras from $E(U)$ to $\mathrm{End}_{\mathbf{R}}(E(U))$. Verify that if $(x_1, \ldots, x_m)$ and $(y_1, \ldots, y_m)$ are m-tuples of elements of U then $(-1)^{m(m-1)/2}\gamma(x_1 \cdots x_m)(y_1 \cdots y_m)$ is equal to the determinant whose entries are the inner products $x_i \cdot y_j$. Show that this gives rise to an inner product for $E(U)$ that extends the given inner product on U and for which the homogeneous components $E^m(U)$ are mutually orthogonal.

Project

Sketch a computer program for accomplishing the following task. Given m elements $u_1, \ldots, u_m$ in $\mathbf{R}^n$, where $n \geqq m$, express the product $u_1 \cdots u_m$ in $E(\mathbf{R}^n)$ as a linear combination of the products $e_{i_1} \cdots e_{i_m}$ with $i_1 < \cdots < i_m$, where $(e_1, \ldots, e_n)$ is the usual basis of $\mathbf{R}^n$.

CHAPTER VI

Polynomials

1. Let R be a ring, and let N^+ be the monoid of the non-negative integers, with addition as the monoid composition. We consider the ring $R[N^+]$ whose elements are the maps from N^+ to R, and whose multiplication is the convolution, as defined in Section II.7. Let x denote the element of $R[N^+]$ that takes the value 1_R (the identity element of R) at the element 1 of N^+ and the value 0_R at every other element of N^+. Then every element f of $R[N^+]$ may be viewed as a formal power series in x, the coefficient of x^n being $f(n)$. More precisely, the purely formal infinite "sum" $\sum_{n \geqq 0} f(n)x^n$ may be regarded as an element of $R[N^+]$ in the evident way, merely by observing that, at each element m of N^+, all the summands $f(n)x^n$ in which n is different from m take the value 0_R, while $f(m)x^m$ takes the value $f(m)$ at m. In this sense, we have

$$f = \sum_{n \geqq 0} f(n)x^n$$

and the multiplication in $R[N^+]$ is determined by R-linearity and the fact that $x^p x^q = x^{p+q}$ for all non-negative exponents p and q. When we have this description in mind, we write $R[[x]]$ for $R[N^+]$, and we refer to it as the *ring of formal power series over* R. Usually, we shall be in a situation where R is commutative, and we note that then $R[[x]]$ is actually an R-algebra, in the sense used before with a field in the role of R.

For every f in $R[N^+]$, let us define the element f' of $R[N^+]$ by $f'(n) = (n + 1)f(n + 1)$. One verifies directly that $(fg)' = f'g + fg'$, and $f' = 0$ if f belongs to R, i.e., if $f(n) = 0$ whenever $n > 0$. Let δ denote the group endomorphism of $R[N^+]$ defined by $\delta(f) = f'$. Then the above properties are expressed by saying that δ is an R-*linear derivation* of $R[N^+]$. Evidently, $\delta(x)$ is the element of $R[N^+]$ that takes the value 1_R at 0 and the value 0_R at every other element of N^+. This is the identity element of $R[N^+]$, which we identify with that of R, and which we write simply as 1 whenever there is no danger of confusion. Thus $\delta(x) = 1$, and now δ appears as the "differentiation with respect to x" in $R[[x]]$.

The *polynomial ring* $R[x]$ is defined as the subring of $R[N^+]$ consisting of the elements that take non-zero values only on a *finite* subset of N^+. These elements, i.e., the *polynomials*, are the linear combinations of finite sequences of powers x^n of x. In fact, if f is a polynomial, then the coefficient of x^n in f equals $f(n)$. The largest n for which $f(n) \neq 0$ is called the *degree* of f.

Observe that $R[x]$ is a very special case of the ring R_H defined in Section V.1. Namely, the case where H is a set with exactly one element, so that every finite sequence of elements of H is determined by its length, i.e., an element of N^+. When R is commutative, the universal mapping property of R_H becomes the following, in our present situation. *For every element u of an arbitrary R-algebra U, there is one and only one homomorphism η_u of R-algebras from $R[x]$ to U such that $\eta_u(x) = u$.*

If f is an element of $R[x]$, let us write $f^\circ(u)$ for $\eta_u(f)$. Then f° is a map from U to U, called the *polynomial function* on U *defined by* f. It is customary to write $f(x)$ for f, and f for f°. However, this notation conflicts with the notation we have used above. The reader may check that, in our notation, if $U = R[x]$ then $f = f^\circ(x)$.

Now let us take R to be a field F. Then the polynomial ring $F[x]$ has good properties with regard to factorization. Indeed, the factorization theory of $F[x]$ is just like that of the ring Z of integers. Our first observation is that *the units of $F[x]$ are precisely the non-zero polynomials of degree* 0, i.e., the elements f such that $f(n) = 0$ if and only if $n > 0$, which we identify with their values $f(0)$ in F. Thus, the units of $F[x]$ are simply the non-zero elements of F. Next, we record the following division algorithm.

Suppose that f and g are polynomials, and that g is of positive degree. There is one and only one pair (q, r) of polynomials such that $f = qg + r$ and the degree of r is strictly smaller than that of g (if $r = 0$, we agree that the degree of r is -1).

In order to see this, let n denote the degree of f, and let d be the degree of g. If $n < d$, then $q = 0$ and $r = f$. Otherwise, note that $f - f(n)g(d)^{-1}x^{n-d}g$ is of degree strictly smaller than n, and that q must be the sum of $f(n)g(d)^{-1}x^{n-d}$ and a polynomial of degree at most $n - d - 1$. Clearly, this establishes the result by induction on n.

From this *Euclidean algorithm*, one derives the decisive fact that *every ideal of $F[x]$ consists simply of the products by arbitrary elements of $F[x]$ of a fixed single element*, i.e., has the form $F[x]f$, where f is a fixed element of $F[x]$. In fact, if the ideal is $\{0\}$ then $f = 0$. Otherwise, f may be taken to be any non-zero element of the ideal having the minimum degree for such elements, and f is determined by the ideal to within non-zero factors from F.

If p and q are polynomials, and $F[x]p + F[x]q = F[x]f$ then f is called a *greatest common divisor* of p and q. By what we have just said, any two greatest common divisors of p and q differ only by a non-zero factor from F. A non-zero polynomial f is called a *prime* element or an *irreducible* element of $F[x]$, if f is not a unit and the only factors of f are the elements αf and α, where α is a unit. As in the case of Z, one proves that *every non-zero element of $F[x]$ is a product of units and prime elements*, and that, *if the ordering of the factors is ignored, such a factorization of an element is determined by that element up to unit factors.*

2. We wish to analyze the ring $\mathbf{R}[x]$ of polynomials with real coefficients in depth. For this, we require the notion of a *continuous function*, which we define in the following general context.

Let S be a set, and suppose S is endowed with a *distance* function, by which is meant a real-valued function D on $S \times S$ satisfying the following conditions.

(1) If p and q are elements of S then $D(p, q) = D(q, p) \geqq 0$, and $D(p, q) = 0$ if and only if $p = q$.
(2) If p, q and r are elements of S then

$$D(p, q) + D(q, r) \geqq D(p, r)$$

We describe this situation by saying that D makes S into a *metric space*. The elements of S are called points, and $D(p, q)$ is called the distance between p and q.

A map f from a metric space S to a metric space T is said to be continuous at a point p of S if, for every positive real number ε, there is a positive real number δ such that the distance between $f(p)$ and $f(q)$ is less than ε for every point q of S whose distance from p is less than δ. We say that f is continuous if it is continuous at every point of S.

Here, we are concerned with the case where S is a closed real interval $[\alpha, \beta]$ and f is a continuous map from S to $\mathbf{R}$. Suppose that f takes on positive values, as well as negative ones. The result we wish to use is that then f must take on the value 0 at some point of $[\alpha, \beta]$.

We prove this by deriving a contradiction from the assumption that f does not take on the value 0 at any point of $[\alpha, \beta]$. Let γ_1 denote the midpoint $(\alpha + \beta)/2$ of our interval. If f had constant sign on each of the intervals $[\alpha, \gamma_1]$ and $[\gamma_1, \beta]$ then f would be of constant sign on $[\alpha, \beta]$, contrary to the initial assumption made on f. Let γ_2 denote the midpoint of one of the intervals $[\alpha, \gamma_1]$ or $[\gamma_1, \beta]$ on which f is not of constant sign, and repeat the argument we have just made. Continuing in this fashion, we generate an infinite sequence $(\gamma_1, \gamma_2, \ldots)$ of points in $[\alpha, \beta]$, such that each γ_n is the midpoint of a subinterval of length $(\beta - \alpha)/2^{n-1}$ on which f changes sign, and the distance between γ_n and γ_{n+1} is equal to $(\beta - \alpha)/2^{n+1}$. In particular, this sequence is a Cauchy sequence whose limit, γ say, belongs to $[\alpha, \beta]$. Moreover, f changes sign between points arbitrarily close to γ. Since f is continuous, this implies that $f(\gamma) = 0$, contrary to the assumption we made for the sake of the argument.

It is clear from the definition of continuity that value-wise sums and products of continuous real-valued functions on a metric space are continuous. In particular, consider the polynomial functions from **R** to **R** corresponding to the elements of $\mathbf{R}[x]$. The polynomial function corresponding to x is the identity map from **R** to **R**, which is evidently continuous. Of course, all constant functions are continuous. It follows that *every polynomial function from* **R** *to* **R** *is continuous.*

An almost immediate consequence of the above results is that *every polynomial (function) with real coefficients that is of odd degree has a real root.* The proof is as follows. Let f be such a polynomial, and denote its degree by $2d + 1$, where d is some non-negative integer. Let γ_n denote the coefficient with which the function corresponding to x^n occurs in f. Then, for every real number α other than 0, we may write the value of f at α in the form

$$f(\alpha) = \alpha^{2d+1}(\gamma_{2d+1} + \gamma_{2d}\alpha^{-1} + \cdots + \gamma_0\alpha^{-2d-1})$$

This shows that, if α is large and positive, then $f(\alpha)$ has the same sign as γ_{2d+1}, and $f(-\alpha)$ has the opposite sign. As we have shown above, it follows that we must have $f(\alpha) = 0$ for some real number α.

A substantial and somewhat devious extension of the above reasoning yields a procedure for locating the real roots of an arbitrary polynomial function on **R**. Suppose that we are given a polynomial function g on **R** of positive degree, and real numbers $\alpha < \beta$ such that $g(\alpha)g(\beta) \neq 0$. A *Sturm sequence* for g and the interval $[\alpha, \beta]$ is a finite sequence $(g_1, \ldots, g_s)$ of polynomial functions on **R** satisfying the following requirements.

(1) If j is an index from $(1, \ldots, s - 1)$, and γ is a root of g_j in $[\alpha, \beta]$, then $g_{j-1}(\gamma)g_{j+1}(\gamma) < 0$, where g_0 stands for g.
(2) For every root γ of g that belongs to $[\alpha, \beta]$, there are real numbers γ_1 and γ_2 such that $g(\rho)g_1(\rho) < 0$ if $\gamma_1 \leqq \rho < \gamma$, while $g(\rho)g_1(\rho) > 0$ if $\gamma < \rho \leqq \gamma_2$.
(3) g_s has no root in $[\alpha, \beta]$.

For a finite sequence $c = (\gamma_0, \ldots, \gamma_m)$ of non-zero real numbers, we define the *number of sign changes* of c as the number of indices i from $(0, \ldots, m - 1)$ for which $\gamma_i\gamma_{i+1} < 0$. If c is an arbitrary finite sequence of real numbers, we define the number of sign changes of c to be that of the sequence obtained from c by deleting the 0's. In the above context of a Sturm sequence, we define $W(\gamma)$, for every real number γ, as the number of sign changes of the sequence $(g(\gamma), g_1(\gamma), \ldots, g_s(\gamma))$.

The importance of a Sturm sequence resides in the fact that, *in the above notation, the number of distinct roots of* g *in* $[\alpha, \beta]$ *is equal to* $W(\alpha) - W(\beta)$.

In order to see this, consider the behavior of $W(\gamma)$ as γ increases from α to β. Evidently, $W(\gamma)$ remains constant while γ ranges over any interval not containing a root of g or any of the g_i's. It follows from (1) that $W(\gamma)$ remains constant while γ passes a root of one or several of the g_i's that is not a root of g (by (3), the case $i = s$ does not arise here). Finally, it follows from (2) that the effect of γ passing a root of g is a decrease of 1 in $W(\gamma)$, due to the change from 1 to 0 of the number of sign changes in $(g(\gamma), g_1(\gamma))$ (appeal to (1) in the case where that root of g is also a root of one or several of the g_i's).

Next, we discuss the construction of Sturm sequences. Let p_0 be any non-zero element of $\mathbf{R}[x]$. If the degree of p_0 is positive, we define p_1 as the formal derivative $\delta(p_0)$. Generally, when p_i and p_{i+1} have been defined as elements of positive degree of $\mathbf{R}[x]$, we use the Euclidean algorithm, with a change of sign in the remainder, to write

$$p_i = q_i p_{i+1} - p_{i+2}$$

thus defining p_{i+2} as an element of $\mathbf{R}[x]$ of degree less than that of p_{i+1}. We continue this construction as long as we obtain non-zero polynomials. The result is a finite sequence $(p_0, \ldots, p_s)$ of non-zero polynomials, and the above recursion shows that each p_i is divisible by p_s, so that $p_i = h_i p_s$, with h_i in $\mathbf{R}[x]$.

Given a non-constant polynomial function f on $\mathbf{R}$, write $f = (p_0)^\circ$, with p_0 in $\mathbf{R}[x]$, and let $(h_0, \ldots, h_s)$ be the sequence obtained from p_0 as just above. Put $g_i = (h_i)^\circ$. Writing g for g_0, we claim that *$(g_1, \ldots, g_s)$ is a Sturm sequence for g in every interval $[\alpha, \beta]$ such that $g(\alpha)g(\beta) \neq 0$.*

Since g_s is the constant 1, requirement (3) for a Sturm sequence is satisfied. Next, for each index j from $(1, \ldots, s-1)$, we have

$$g_{j-1} = (q_j)^\circ g_j - g_{j+1}$$

This shows that a common root of g_j and $g_{j-1}g_{j+1}$ must be a root of each g_i, contradicting $g_s = 1$. Therefore, g_j and $g_{j-1}g_{j+1}$ have no common root. With this information, the above recursion shows that requirement (1) for a Sturm sequence is satisfied.

Now suppose that γ is a root of g. Then γ is also a root of f. Since $f = (p_0)^\circ$, the Euclidean algorithm for the pair $(p_0, x - \gamma)$ shows that $p_0 = (x-\gamma)^e h$, where e is some *positive* exponent, and h is some element of $\mathbf{R}[x]$ that is no longer divisible by $x - \gamma$, so that $h^\circ(\gamma) \neq 0$. It follows that

$$p_1 = e(x-\gamma)^{e-1}h + (x-\gamma)^e \delta(h)$$

Thus, the highest power of $x - \gamma$ dividing both p_0 and p_1 is $(x-\gamma)^{e-1}$, and the recursion for the p_i's shows that p_s is divisible by $(x-\gamma)^{e-1}$. Consequently, h_1 is not divisible by $x - \gamma$, which means that γ is not a root of g_1. For every real number ρ, we have

$$(p_0)^\circ(\rho) = (p_s)^\circ(\rho)g(\rho) \quad \text{and} \quad (p_1)^\circ(\rho) = (p_s)^\circ(\rho)g_1(\rho)$$

Multiplying these together and using the above expressions for p_0 and p_1, we obtain

$$(p_s)^\circ(\rho)^2 g(\rho) g_1(\rho) = (\rho - \gamma)^{2e-1}[eh^\circ(\rho)^2 + (\rho - \gamma)(h\delta(h))^\circ(\rho)]$$

Taking ρ near γ and recalling that $h^\circ(\gamma) \neq 0$, we see from this that the pair (g, g_1) satisfies requirement (2) for a Sturm sequence.

Now consider an interval $[\alpha, \beta]$, where $f(\alpha)f(\beta) \neq 0$, so that we have also $g(\alpha)g(\beta) \neq 0$. Let $W(\gamma)$ denote the number of sign changes in the sequence $(g(\gamma), g_1(\gamma), \ldots, g_s(\gamma))$. From the above, we conclude that the number of distinct roots of g in $[\alpha, \beta]$ is equal to $W(\alpha) - W(\beta)$. We know that every root of g is also a root of f. Conversely, if γ is a root of f, we know from the above that $p_0 = (x - \gamma)^e h$, while the highest power of $x - \gamma$ dividing p_s is $(x - \gamma)^{e-1}$. Consequently, γ is also a root of g. Thus, the set of roots of g coincides with the set of roots of f.

Now let $f_i = (p_i)^\circ$, and let $V(\gamma)$ denote the number of sign changes in the sequence $(f(\gamma), f_1(\gamma), \ldots, f_s(\gamma))$. We have $f_i(\gamma) = (p_s)^\circ(\gamma)g_i(\gamma)$ for each index i, including 0, if we agree that f_0 stands for f. Since $f(\alpha)f(\beta) \neq 0$, we have also $(p_s)^\circ(\alpha)(p_s)^\circ(\beta) \neq 0$. Therefore, $V(\alpha) = W(\alpha)$ and $V(\beta) = W(\beta)$. Thus, we have the following result (Sturm's theorem).

Let f be a polynomial function on $\mathbf{R}$ having positive degree, and let $(f_1, \ldots, f_s)$ be the sequence of polynomial functions obtained as above from the Euclidean algorithm, starting from the polynomial representing f and the formal derivative of that polynomial. Let $V(\gamma)$ denote the number of sign changes in the sequence $(f(\gamma), f_1(\gamma), \ldots, f_s(\gamma))$, and let $[\alpha, \beta]$ be an interval such that $f(\alpha)f(\beta) \neq 0$. Then the number of distinct roots of f in $[\alpha, \beta]$ is equal to $V(\alpha) - V(\beta)$.

3. We wish to establish the fundamental fact that *every polynomial function on* $\mathbf{C}$ *of positive degree has a root in* $\mathbf{C}$. We denote the identity function on $\mathbf{C}$ by z, so that the polynomial functions on $\mathbf{C}$ are the linear combinations with coefficients in $\mathbf{C}$ of powers z^n of z.

First, we deal with the simplest non-trivial case of a polynomial function $z^2 - \gamma$, where γ is an arbitrary complex number. We write $\gamma = \alpha + \beta i$, where α and β are real numbers. The conditions that a complex number $\rho + \sigma i$, with ρ and σ real, must satisfy in order to be a root of our polynomial are

$$\rho^2 - \sigma^2 = \alpha \quad \text{and} \quad 2\rho\sigma = \beta$$

The first of these conditions gives the condition

$$4\rho^2\sigma^2 - 4\sigma^4 = 4\alpha\sigma^2$$

Together with the second condition, this yields the condition

$$(\sigma^2 + \alpha/2)^2 = (\alpha^2 + \beta^2)/4$$

Consider the polynomial function f on $\mathbf{R}$ that is defined by

$$f(\tau) = \tau^2 - (\alpha^2 + \beta^2)/4$$

Since $f(0) \leqq 0$, while $f(\tau) > 0$ for large positive real numbers τ, it follows from the continuity of f that there is a non-negative real number τ such that $\tau^2 = (\alpha^2 + \beta^2)/4$. Now $\tau - \alpha/2$ is non-negative, because τ is non-negative and $\tau^2 \geqq \alpha^2/4$. Therefore, we can make the same continuity argument as just above to conclude that there is a non-negative real number σ such that $\sigma^2 = \tau - \alpha/2$. Now we have

$$\sigma^2 + \alpha = \tau + \alpha/2 \geqq 0$$

Using the same continuity argument once more, we conclude that there is a non-negative real number μ such that $\mu^2 = \sigma^2 + \alpha$. If $\beta < 0$, put $\rho = -\mu$. Otherwise, put $\rho = \mu$. Now one can verify directly that $2\rho\sigma = \beta$, and hence that

$$(\rho + \sigma i)^2 = \alpha + \beta i$$

Our next step is the proof that, for every complex number γ and every positive exponent n, there is a complex number δ such that $\delta^n = \gamma$. Since we have just established the existence of square roots of arbitrary complex numbers, it suffices to deal with the case where n is odd.

If γ is real, we already know that there is a real solution δ, because we have shown in Section 2 that every real polynomial of odd degree has a root in $\mathbf{R}$. Therefore, we assume from here on that γ does not coincide with its conjugate γ^*. Now we observe that it suffices to find a complex number σ satisfying $(\sigma\sigma^*)^n = \gamma\gamma^*$ and $\sigma^n\gamma^* = (\sigma^*)^n\gamma$. Indeed, for such a σ, we have

$$\sigma^{2n}\gamma^* = \sigma^n(\sigma^*)^n\gamma = \gamma\gamma^*\gamma$$

from which it evidently follows that $\sigma^{2n} = \gamma^2$. Hence, σ^n is one of γ or $-\gamma$. Accordingly, we set δ equal to σ or $-\sigma$. Since n is odd, this makes $\delta^n = \gamma$.

Now we try to find such a σ in the form $\beta(\alpha + i)$, with α and β real. Then

$$\begin{aligned} \sigma\sigma^* &= \beta^2(\alpha^2 + 1) \quad \text{and} \\ \sigma^n\gamma^* - (\sigma^*)^n\gamma &= \beta^n[(\alpha + i)^n\gamma^* - (\alpha - i)^n\gamma] \\ &= \beta^n i[\alpha^n(\gamma^* - \gamma)/i + \alpha^{n-1}\mu_1 + \cdots + \mu_n] \end{aligned}$$

where the μ_j's are certain *real* numbers depending only on γ, not on α. Since $(\gamma^* - \gamma)/i$ is a non-zero real number, the last expression in square brackets is of the form $f(\alpha)$, where f is a real polynomial function of degree n. Since n is odd, f has a real root. Accordingly, we take our above real number α such that $f(\alpha) = 0$. Then we have $\sigma^n\gamma^* = (\sigma^*)^n\gamma$ for every choice of the real number β. It remains to be shown that β can be chosen so that $(\sigma\sigma^*)^n = \gamma\gamma^*$, i.e., $\beta^{2n}(\alpha^2 + 1)^n = \gamma\gamma^*$. Since $\gamma\gamma^*$ is positive, we see as above for square roots that there is a positive real number τ such that $\tau^n = \gamma\gamma^*$. Now it suffices for β to satisfy $\beta^2 = \tau/(\alpha^2 + 1)$, so that a suitable β can indeed be found.

In order to deal with general polynomial functions on **C**, we require the following fact from real analysis. *Suppose that S is a non-empty set of real numbers having a lower bound. Then S has a greatest lower bound, i.e., a lower bound σ such that $\sigma \geqq \tau$ for every lower bound τ of S in* **R**.

This can be proved as follows. Let ρ_0 be a lower bound of S, and let σ_0 be any element of S. Put $\tau_0 = (\rho_0 + \sigma_0)/2$. If there is an element of S in $[\rho_0, \tau_0]$, let σ_1 be such an element, and put $\rho_1 = \rho_0$. Otherwise, put $\sigma_1 = \sigma_0$ and $\rho_1 = \tau_0$. It is easy to see that iteration of this construction yields a Cauchy sequence $(\rho_0, \rho_1, \ldots)$ of lower bounds of S such that, for each index n, there is a point of S whose distance from ρ_n is no greater than $(\sigma_0 - \rho_0)/2^n$. Evidently, the limit of this Cauchy sequence is the required greatest lower bound of S.

Now let n be a positive integer, and consider a polynomial function

$$f = z^n + \alpha_1 z^{n-1} + \cdots + \alpha_n$$

on **C**, where each α_j is a complex number. We shall prove that there is a complex number γ such that $f(\gamma) = 0$.

For every complex number τ, we denote the *absolute value* of τ by $|\tau|$. This is the non-negative real number whose square is equal to $\tau\tau^*$, and it is the distance of τ from 0 when **C** is regarded as $\mathbf{R}^2$. If α is a non-zero complex number, we may write

$$f(\alpha) = \alpha^n[1 + (\alpha_1/\alpha) + \cdots + (\alpha_n/\alpha^n)]$$

This shows that $|f(\alpha)|/|\alpha|^n$ approaches 1 as $|\alpha|$ is made large. Consequently, there is a positive real number η such that $|f(\alpha)| > |f(0)|$ whenever $|\alpha| > \eta$. Let μ denote the greatest lower bound of the set of non-negative real numbers $|f(\alpha)|$, with α in **C**. First, we show that there is a complex number γ such that $|f(\gamma)| = \mu$.

Regarding **C** as the Euclidean plane $\mathbf{R}^2$, let A_0 be the square in **C** with center 0 and sides of length 2η, parallel to the usual coordinate axes. Clearly, the greatest lower bound of the set of real numbers $|f(\alpha)|$, with α in A_0, is still equal to μ. For every non-negative integer m, choose a point β_m in A_0 such that $|f(\beta_m)| < \mu + 2^{-m}$. Now partition A_0 into 4 congruent subsquares by drawing the line segments joining the midpoints of opposite sides. At least one of these subsquares, say A_1, contains β_m for infinitely many m's. Put $\gamma_0 = \beta_0$, and let γ_1 be one of the β_m's in A_1, with $m \geqq 1$, so that

$$|f(\gamma_1)| < \mu + 2^{-1}.$$

Iterating this selection process, we obtain an infinite sequence $(\gamma_0, \gamma_1, \ldots)$ such that $|f(\gamma_m)| < \mu + 2^{-m}$ and $|\gamma_{m+1} - \gamma_m| < \eta\sqrt{2}\, 2^{1-m}$. Evidently, the coordinates of these points γ_m constitute Cauchy sequences of real numbers, and we may conclude that the sequence of γ_m's has a limit, γ say. It is clear from the construction of the sequence of γ_m's that we must have $|f(\gamma)| = \mu$.

Now we show that $\mu = 0$. Clearly, there are complex numbers $\delta_1, \ldots, \delta_{n-1}$, depending only on γ and the coefficients of f, such that

$$f(\gamma + \alpha) = f(\gamma) + \delta_1\alpha + \cdots + \delta_{n-1}\alpha^{n-1} + \delta_n\alpha^n$$

for every complex number α, where $\delta_n = 1$. Let k be the lowest index with $\delta_k \neq 0$. Then we have

$$f(\gamma + \alpha) = f(\gamma) + \alpha^k(\delta_k + \delta_{k+1}\alpha + \cdots + \delta_n\alpha^{n-k})$$

There is a positive real number ε such that

$$|\delta_{k+1}\alpha + \cdots + \delta_n\alpha^{n-k}| < |\delta_k|/2$$

whenever $|\alpha| < \varepsilon$. Choose a real number δ such that $0 < \delta < 1$ and $\delta\mu < \varepsilon^k|\delta_k|$. We know from what we have already shown that there is a complex number τ such that

$$\tau^k = -\delta f(\gamma)/\delta_k$$

This implies that $|\tau|^k = \delta\mu/|\delta_k| < \varepsilon^k$, whence $|\tau| < \varepsilon$. Now we have

$$\begin{aligned} f(\gamma + \tau) &= f(\gamma) - (\delta f(\gamma)/\delta_k)[\delta_k + \delta_{k+1}\tau + \cdots + \delta_n\tau^{n-k}] \\ &= (1 - \delta)f(\gamma) - (\delta f(\gamma)/\delta_k)[\delta_{k+1}\tau + \cdots + \delta_n\tau^{n-k}] \end{aligned}$$

The last expression in square brackets is of absolute value less than $|\delta_k|/2$. Consequently, we have

$$|f(\gamma + \tau)| \leqq (1 - \delta)\mu + \delta\mu/2 = (1 - \delta/2)\mu$$

which contradicts the definition of μ, unless $\mu = 0$. Our conclusion is that $f(\gamma) = 0$, so that the statement at the beginning of this section has been established.

It is an almost immediate consequence of what we have just proved that *the prime elements of the polynomial ring* $\mathbf{C}[x]$ *are simply the polynomials of degree* 1. Next, for every f in $\mathbf{C}[x]$, let f^* denote the element of $\mathbf{C}[x]$ that is defined by $f^*(n) = f(n)^*$ for every n in N^+. In other words, f^* is the polynomial whose coefficient are the complex conjugates of the coefficients of f. Evidently, the map sending each f to f^* is a ring automorphism of $\mathbf{C}[x]$, i.e., a ring isomorphism from $\mathbf{C}[x]$ to $\mathbf{C}[x]$. The elements f such that $f = f^*$ constitute precisely the subring $\mathbf{R}[x]$ of $\mathbf{C}[x]$. Now let f be an element of $\mathbf{R}[x]$. Apart from unit factors, the irreducible factors of f in $\mathbf{C}[x]$ are of the form $x + \alpha$, with α in $\mathbf{C}$. Since $f = f^*$, it is clear that $x + \alpha^*$ is a factor of f whenever $x + \alpha$ is a factor of f. Hence, if α is not real then $(x + \alpha)(x + \alpha^*)$ is a factor of f, and this factor belongs to $\mathbf{R}[x]$. Pairing up the irreducible non-real factors of f in this way, we see that, up to unit factors, the irreducible factors of f in $\mathbf{R}[x]$ are either of the form $x + \alpha$, with α in $\mathbf{R}$, or of the form $(x + \alpha)(x + \alpha^*)$, with α in $\mathbf{C}$ but not in $\mathbf{R}$. The second type can be written in the form $(x + \rho)^2 + \sigma^2$, where ρ and σ are real numbers, and $\sigma \neq 0$. The general conclusion is that *the irreducible elements of* $\mathbf{R}[x]$ *are either of degree* 1, *or non-zero real multiples of polynomials of the form* $(x + \rho)^2 + \sigma^2$, *where* ρ *and* σ *are real numbers and* $\sigma \neq 0$. *Conversely, all polynomials of this type are irreducible in* $\mathbf{R}[x]$.

4. We shall discuss approximation to continuous functions by polynomial functions. Fix an interval $[\alpha, \beta]$ on $\mathbf{R}$, and let f be a continuous real-valued function on $[\alpha, \beta]$. Using a partitioning argument like the one near the beginning of Section 2, we show that the set of values of f is bounded. Indeed, suppose that this is not the case. Then we construct a sequence of points γ_n in $[\alpha, \beta]$ such that, for each n, the distance between γ_n and γ_{n+1} is equal to $(\beta - \alpha)/2^{n+1}$, and each γ_n is the midpoint of a subinterval of $[\alpha, \beta]$ that is of length $(\beta - \alpha)/2^n$ and on which f is not bounded. If γ is the limit of this Cauchy sequence $(\gamma_0, \gamma_1, \ldots)$, then γ belongs to $[\alpha, \beta]$ and, for every positive real number ε, there is a subinterval of $[\alpha, \beta]$ whose length is less than ε, which contains γ and on which f is not bounded. Clearly, this contradicts the continuity of f. Our conclusion is that *every continuous real-valued function on a finite closed real interval is bounded.*

For every f as above, we denote the least upper bound of the set of absolute values $|f(\gamma)|$, with γ ranging over $[\alpha, \beta]$, by $\|f\|$. We call this the *norm* of f. This norm function on the $\mathbf{R}$-space of all continuous real-valued functions on $[\alpha, \beta]$ obeys the same formal rules as the Euclidean norm $(p \cdot p)^{1/2}$ in $\mathbf{R}^n$, i.e., we have

$$\|f + g\| \leqq \|f\| + \|g\| \quad \text{and} \quad \|\rho f\| = |\rho| \, \|f\|$$

for all elements f and g of our $\mathbf{R}$-space of functions and for every real number ρ. In particular, this norm makes our space of functions into a metric space, the distance between two functions f and g being defined as $\|f - g\|$.

A partitioning and construction of a Cauchy sequence of the type we have used several times by now shows that there is a point γ in $[\alpha, \beta]$ such that $\|f\| = |f(\gamma)|$. In other words, *our norm is actually the maximum absolute value, for every continuous real-valued function on a closed real interval.*

Let t stand for the identity map on $[\alpha, \beta]$, so that the restrictions to $[\alpha, \beta]$ of the polynomial functions on $\mathbf{R}$ are the $\mathbf{R}$-linear combinations of the powers of t. We fix a continuous function f on $[\alpha, \beta]$, as well as a non-negative integer n, and we consider the problem of approximating f by a polynomial function of degree $\leqq n$. Accordingly, we define a real-valued function D on $\mathbf{R}^{n+1}$ by the formula

$$D(\mu_0, \ldots, \mu_n) = \left\| f - \sum_{i=0}^{n} \mu_i t^i \right\|$$

From the properties of the norm noted above, we obtain

$$\begin{aligned}|D(\mu_0+\delta_0,\ldots,\mu_n+\delta_n)-D(\mu_0,\ldots,\mu_n)| &\leqq \left\|\sum_{i=0}^{n}\delta_i t^i\right\| \\ &\leqq \sum_{i=0}^{n}|\delta_i|\cdot\|t^i\| \\ &\leqq M\sum_{i=0}^{n}|\delta_i|\end{aligned}$$

where M stands for the largest of the norms $\|t^i\|$ for $i = 0, \ldots, n$. It is clear from this that D is a continuous function.

Let B denote the hollow cube in $\mathbf{R}^{n+1}$ consisting of all points $(\mu_0, \ldots, \mu_n)$ for which the maximum of $|\mu_0|, \ldots, |\mu_n|$ is precisely equal to 1. Let E denote the function on B defined by

$$E(\mu_0,\ldots,\mu_n) = \left\|\sum_{i=0}^{n}\mu_i t^i\right\|$$

Like the function D, this function E is continuous. Now note that B is the union of 2^{n+1} faces B_{i+} and B_{i-}, with $i = 0, 1, \ldots, n$, where B_{i+} is defined by the restriction $\mu_i = 1$, and B_{i-} is defined by $\mu_i = -1$. On each of these faces, the coordinates μ_j with $j \neq i$ range freely over the interval $[-1, 1]$. By inserting the appropriate integer multiples of $1/2^k$, with $k = 0, 1, \ldots$, in this interval, we may partition each B_i into a union of n-dimensional solid cubes of sidelength $1/2^k$. Using such a partitioning, we may proceed as in Section 3 (when we showed that there is a point in $\mathbf{C}$ at which a given polynomial has the minimum absolute value) and show that E attains a minimum at some point of B. This minimum is the norm of a non-zero polynomial function on $[\alpha, \beta]$, so that it is a *positive* real number, L say.

Now we are in a position to confine the search for a point of $\mathbf{R}^{n+1}$ at which the value of D is minimal to some solid cube in $\mathbf{R}^{n+1}$, as follows. We have

$$D(\mu_0,\ldots,\mu_n) \geqq \left\|\sum_{i=0}^{n}\mu_i t^i\right\| - \|f\|$$

Let μ denote the maximum $|\mu_i|$ for $i = 0, 1, \ldots, n$ here, and assume $\mu > 0$. Then we may write

$$\left\|\sum_{i=0}^{n}\mu_i t^i\right\| = \mu\left\|\sum_{i=0}^{n}(\mu_i/\mu)t^i\right\| = \mu E(\mu_0/\mu,\ldots,\mu_n/\mu)$$

which shows that we have, actually for all points $(\mu_0, \ldots, \mu_n)$,

$$\left\|\sum_{i=0}^{n}\mu_i t^i\right\| \geqq \mu L$$

Now the above inequality for D shows that if $\mu > 2\|f\|/L$ then

$$D(\mu_0, \ldots, \mu_n) > \|f\| = D(0, \ldots, 0).$$

Consequently, the greatest lower bound of the set of all values of D coincides with the greatest lower bound of the set of values of the restriction of D to the solid cube consisting of the points $(\mu_0, \ldots, \mu_n)$ with $|\mu_i| \leqq 2\|f\|/L$ for each i. To this restriction of D, we can apply the argument used above for E to conclude that D attains a minimum. This is the crude part of the following result (which is due to Chebyshev).

Let f be a continuous real-valued function on a real interval $[\alpha, \beta]$. Among the polynomial functions on $[\alpha, \beta]$ of degree $\leqq n$, there is exactly one, p say, for which $\|f - p\|$ is minimal. Moreover, there are $n + 2$ points γ_i, with $\alpha \leqq \gamma_0 < \cdots < \gamma_{n+1} \leqq \beta$, such that $|f(\gamma_i) - p(\gamma_i)| = \|f - p\|$, and the sign of $f(\gamma_i) - p(\gamma_i)$ alternates as i goes from 0 to $n + 1$ (unless $p = f$).

We have just shown that there is at least one polynomial function p of degree n such that $\|f - p\|$ is minimal. If $\|f - p\| = 0$ then there is no more to be proved. Therefore, we assume that $\|f - p\| > 0$. We have seen above that the norm of a continuous function is actually the maximum absolute value. Accordingly, let γ_0 be the smallest real number in $[\alpha, \beta]$ for which $|f(\gamma_0) - p(\gamma_0)| = \|f - p\|$. Replacing f with $-f$ and p with $-p$, if necessary, we arrange to have $f(\gamma_0) - p(\gamma_0) = \|f - p\|$. If there were no point ρ in $[\alpha, \beta]$ such that $f(\rho) - p(\rho) = -\|f - p\|$ then, for a sufficiently small positive real constant η, we would have $\|f - (p + \eta)\| < \|f - p\|$, contradicting the choice of p. Accordingly, let γ_1 be the smallest such ρ. Then we have $\gamma_1 > \gamma_0$ and $f(\gamma_1) - p(\gamma_1) = -\|f - p\|$.

Generally, suppose that we have already found

$$\alpha \leqq \gamma_0 < \gamma_1 < \cdots < \gamma_k \leqq \beta$$

such that $f(\gamma_i) - p(\gamma_i) = (-1)^i\|f - p\|$ for each i. If the set of points ρ in $[\alpha, \beta]$ such that $\rho > \gamma_k$ and $f(\rho) - p(\rho) = (-1)^{k+1}\|f - p\|$ is non-void, we define γ_{k+1} as the smallest such ρ. We shall obtain a contradiction from the assumption that this construction terminates with $k < n + 1$.

Choose real numbers $\delta_0, \delta_1, \ldots, \delta_{k-1}$ such that $\gamma_i < \delta_i < \gamma_{i+1}$ for each i, and put

$$g = (-1)^k(t - \delta_0)\cdots(t - \delta_{k-1})$$

Then each of the intervals $[\alpha, \delta_0)$, $(\delta_0, \delta_1), \ldots, (\delta_{k-2}, \delta_{k-1})$, $(\delta_{k-1}, \beta]$ contains exactly one of the γ_i's. Here, the use of a round parenthesis in the place of a square bracket indicates that the endpoint next to it is to be omitted from the interval. At each point of the interval containing γ_i, the sign of the value of g is $(-1)^i$. It is easy to see from this that, for a sufficiently small positive real number η, one has $\|f - (p + \eta g)\| < \|f - p\|$. Since $k < n + 1$, the degree k of the polynomial function g is at most n, so that $p + \eta g$ is still a polynomial function of degree $\leqq n$, and our inequality contradicts the choice of p.

Our conclusion is that the above construction of γ_i's does not terminate before the index $n + 1$ is reached, so that the existence of points $\gamma_0, \ldots, \gamma_{n+1}$ as described in the above main statement is established.

Now suppose that q is a polynomial function on $[\alpha, \beta]$ of degree $\leqq n$ that is also a best approximation to f, in the sense that $\|f - q\| = \|f - p\|$. Then we have

$$\|f - (p + q)/2\| \leqq (1/2)\|f - p\| + (1/2)\|f - q\| = \|f - p\|$$

By what we have already proved, there are $n + 2$ distinct points γ in $[\alpha, \beta]$ such that $|f(\gamma) - (1/2)(p(\gamma) + q(\gamma))| = \|f - p\|$. The expression on the left is at most equal to $(1/2)|f(\gamma) - p(\gamma)| + (1/2)|f(\gamma) - q(\gamma)|$, and each of $|f(\gamma) - p(\gamma)|$ and $|f(\gamma) - q(\gamma)|$ is at most equal to $\|f - p\|$. It follows that the equalities must hold, and also that $f(\gamma) - p(\gamma)$ and $f(\gamma) - q(\gamma)$ are of the same sign. Thus, $f(\gamma) - p(\gamma) = f(\gamma) - q(\gamma)$, and therefore $p(\gamma) = q(\gamma)$, for $n + 2$ distinct points γ. This implies that $p = q$, because the polynomial function $p - q$ of degree $\leqq n$ has more than n roots only if it is the zero function.

The existence of the points γ_i is of interest chiefly because it *is also a sufficient condition for p to be a best possible approximation to f among all polynomial functions of degree $\leqq n$.* In fact, suppose that f is an arbitrary real-valued function on $[\alpha, \beta]$, that p is a polynomial function of degree $\leqq n$, and that there are points γ_i such that $\alpha \leqq \gamma_0 < \cdots < \gamma_{n+1} \leqq \beta$ and $(-1)^i(f(\gamma_i) - p(\gamma_i)) > 0$ for each i. Then, for every polynomial function q of degree $\leqq n$, there is at least one index j for which $|f(\gamma_j) - q(\gamma_j)|$ is no smaller than the minimum of the set of positive real numbers $(-1)^i(f(\gamma_i) - p(\gamma_i))$. The reason for this is that otherwise one has

$$\begin{aligned}(-1)^i(q(\gamma_i) - p(\gamma_i)) &= (-1)^i(f(\gamma_i) - p(\gamma_i)) - (-1)^i(f(\gamma_i) - q(\gamma_i)) \\ &> 0 \quad \text{for each index } i.\end{aligned}$$

Thus, $q - p$ has at least $n + 1$ changes of sign, and hence at least $n + 1$ roots. Since it is a polynomial function of degree at most n, we have the contradiction $p = q$.

5. Let F be a field. We are interested in the use of polynomials for finding solutions of recursion equations in F, of the simplest type. Our objects of study are sequences of elements of F, which we shall view quite formally as maps from the set N of all non-negative integers to F. The most frequently occurring sequences are the *polynomial sequences*, by which we mean functions p from N to F given by a polynomial formula

$$p(k) = \sum_{i=0}^{n} k^i \alpha_i$$

where the α_i's are elements of F, and where the multiplication by an element of N is interpreted as iterated addition. Moreover, in the above formula, k^0 is to be interpreted as 1, even for $k = 0$. With the value-wise operations, the polynomial sequences in F constitute a commutative F-algebra, P say.

The other sequences with which we shall be concerned are the *exponential sequences*, i.e., the maps g from N to F given by $g(k) = \mu^k$, where μ is some fixed non-zero element of F. The most general sequences under discussion will be the *elementary sequences*, by which we mean the linear combinations, with coefficients in P, of exponential sequences. These constitute a commutative P-algebra, which we denote by E. Thus, the elements of E are of the form $\sum_{i=0}^{n} p_i g_i$, where the p_i's are polynomial sequences and the g_i's are exponential sequences. If g_i is given by $g_i(k) = \mu_i^k$, with μ_i in F, then the value at k of the elementary sequence exhibited above is $\sum_{i=0}^{n} p_i(k)\mu_i^k$.

The first result we wish to prove is that *if F is any field containing the field* $\mathbf{Q}$ *of rational numbers as its smallest subfield then the set of exponential sequences is linearly independent over the ring P of polynomial sequences.* In other words, we wish to prove that a sequence $\sum_{i=0}^{r} p_i g_i$ as above is the 0-sequence only if each of the polynomial sequences p_i is the 0-sequence. The assumption that F contains $\mathbf{Q}$ really means that $n\alpha \neq 0$ whenever α is a non-zero element of F and n is a positive integer, so that every element n of N may be identified with the corresponding element $n1_F$ of F.

In proving this result, we use the following two operations on E. The first is the *difference operator* τ, which is defined by $\tau(f)(k) = f(k+1) - f(k)$ for every element k of N. The second is the *shift operator* σ, which is defined by $\sigma(f)(k) = f(k+1)$. Evidently, τ and σ are group endomorphisms of E, with respect to addition. The behavior of τ with respect to the multiplication in E is that of differentiation, *twisted* by the intervention of σ. The formula is

$$\tau(fg) = \tau(f)g + \sigma(f)\tau(g)$$

for all elements f and g of E.

Suppose that $g_0, \ldots, g_r$ are distinct exponential sequences, and that $p_0, \ldots, p_r$ are polynomial sequences such that $\sum_{i=0}^{r} p_i g_i = 0$. We proceed by induction on r in order to prove that each $p_i = 0$. The conclusion evidently holds when $r = 0$. Therefore, we assume that $r > 0$, and that the result has been established in the lower cases. Suppose that $g_i(k) = \mu_i^k$ for every k in N, and define the exponential sequence h_i by $h_i(k) = (\mu_i/\mu_0)^k$. Then

$$p_0 + \sum_{i=1}^{r} p_i h_i = 0$$

We have $\tau(h_i) = \gamma_i h_i$, where γ_i is the element $(\mu_i/\mu_0) - 1$ of F, so that $\gamma_i \neq 0$ for $i > 0$. Since p_0 is a polynomial sequence, there is a positive integer e such that $\tau^e(p_0) = 0$. Therefore, our above equation yields

$$\sum_{i=1}^{r} p_{i,e} h_i = 0$$

where the polynomial sequences $p_{i,e}$ are given recursively by $p_{i,0} = p_i$ and $p_{i,s+1} = \tau(p_{i,s}) + \sigma(p_{i,s})\gamma_i$, as follows from the formal property of τ noted above.

Our inductive hypothesis yields $p_{i,e} = 0$ for each $i > 0$. Now suppose that it has already been shown, for some s with $0 \leqq s < e$, that $p_{i,s+1} = 0$. By virtue of the above recursion, this means that $\tau(p_{i,s}) + \sigma(p_{i,s})\gamma_i = 0$, i.e.,

$$(1 + \gamma_i)p_{i,s}(k + 1) - p_{i,s}(k) = 0$$

for every k in N. If $p_{i,s} \neq 0$, and if ρ denotes the highest coefficient in the defining formula for $p_{i,s}$, then the coefficient of the highest power of k occurring on the left in the above equation is equal to $\gamma_i\rho$. Thus, we must have $\gamma_i\rho = 0$. Since $\gamma_i \neq 0$ for each i other than 0, it follows that $p_{i,s} = 0$. Thus our conclusion is that $p_i = 0$ for each $i > 0$, and the original relation now shows that also $p_0 = 0$.

Now we are ready to deal with the following problem. We are given a finite sequence $(\alpha_1, \ldots, \alpha_k)$ of elements of F, with $\alpha_k \neq 0$, and we wish to find the sequences f in F satisfying the recursion

$$f(n + k) + \alpha_1 f(n + k - 1) + \cdots + \alpha_k f(n) = 0$$

This may be written in the form

$$(\sigma^k + \alpha_1\sigma^{k-1} + \cdots + \alpha_k I)(f) = 0$$

where I stands for the identity map on the set of all sequences in F. Accordingly, we should consider the polynomial

$$q = x^k + \alpha_1 x^{k-1} + \cdots + \alpha_k$$

in $F[x]$. Suppose that q has a root μ in F. Then q is divisible by $x - \mu$ in $F[x]$, and we simplify our problem as follows. Let e be any positive exponent such that q is divisible by $(x - \mu)^e$. Then we may write

$$\sigma^k + \alpha_1\sigma^{k-1} + \cdots + \alpha_k I = J \circ (\sigma - \mu I)^e$$

where J is the map on the set of sequences corresponding to the quotient of q by $(x - \mu)^e$ in the evident way. Therefore, every sequence f that is annihilated by $(\sigma - \mu I)^e$ is a solution of our recursion.

Since $\alpha_k \neq 0$, we have $\mu \neq 0$, and we let g_μ denote the exponential sequence determined by μ, so that $g_\mu(n) = \mu^n$. We claim that, *if p is any polynomial sequence of formal degree less than e, then pg_μ is a solution of our recursion.* In order to see this, we merely have to observe that

$$(\sigma - \mu I)(pg_\mu) = \mu(\sigma(p) - p)g_\mu$$

and that $\sigma(p) - p$ is a polynomial sequence whose formal degree is smaller than that of p, so that it follows inductively that $(\sigma - \mu I)^e$ annihilates pg_μ.

Now suppose that we have a complete factorization

$$q = (x - \mu_1)^{e_1} \cdots (x - \mu_r)^{e_r}$$

where the μ_i's are pairwise distinct non-zero elements of F, and the e_i's are positive exponents. The degree k of q is equal to $\sum_{i=1}^{r} e_i$. For each i, we have e_i solutions $f_{i,s}$ of our recursion, where

$$f_{i,s}(n) = n^s \mu_i^n$$

and where s ranges over the set of non-negative integers less than e_i. Clearly, the solutions f of our recursion constitute a vector space over F, because every F-linear combination of solutions is still a solution. Since the values $f(0), \ldots, f(k-1)$ of a solution f determine the solution completely, via the recursion, it is clear that the dimension of the space of solutions is at most equal to k. If F contains $\mathbf{Q}$, then we know from the beginning of this section that the exponential sequences are linearly independent over the ring of polynomial sequences. It is clear from this that the above set of k solutions $f_{i,s}$ is linearly independent over F. Thus, *if F contains $\mathbf{Q}$ then the above solutions $f_{i,s}$ constitute an F-basis of the space of all solutions of the given recursion.*

Exercises

1. Let F be a field, $p = \alpha_0 + \alpha_1 x + \cdots + \alpha_n x^n$ a polynomial in $F[x]$. Using the notation of Section 1, let U be an F-algebra, and let p° denote the polynomial map from U to U defined by p. Let δ denote the differentiation with respect to x in $F[x]$, and let u be an element of U. Define a doubly indexed sequence of elements $u_{e,i}$ in U, where $e = 0, \ldots, n$ and the accompanying indices i range from 0 up to $n - e$, by the following recursion.

$$\begin{aligned} u_{e,0} &= \alpha_n && \text{for every } e \\ u_{0,i} &= u u_{0,i-1} + \alpha_{n-i} && \text{for } i > 0 \\ u_{e,i} &= u u_{e,i-1} + u_{e-1,i} && \text{for } e > 0 \quad \text{and} \quad i > 0 \end{aligned}$$

Show that, for every e, one has $u_{e,n-e} = \left(\frac{1}{e!}\delta^e\right)(p)^\circ(u)$ (note that the endomorphism $\frac{1}{e!}\delta^e$ of $F[x]$ does not really involve a division: it is the linear endomorphism sending each x^m to the sum of $\binom{m}{e}$ summands x^{m-e}).

2. With the help of the end of Section V.2, show that every linear endomorphism of $\mathbf{R}^3$ has a characteristic vector. Apply this result to show that every orthogonal linear transformation of $\mathbf{R}^3$ whose determinant is positive acts as a rotation in some plane containing the origin and leaves the points of the orthogonal complement of that plane fixed.

3. Let $\pi_0, \ldots, \pi_n$ be pairwise distinct elements of a field F, and let $(\alpha_0, \ldots, \alpha_n)$ be an arbitrary sequence of elements of F. Show that there is one and only one polynomial function f of degree $\leqq n$ on F such that $f(\pi_i) = \alpha_i$ for each i.

4. Let ρ and σ be elements of a finite commutative group G, let r and s be the orders of ρ and σ, and let m denote the least common multiple of r and s. Using factorization into products of primes, show that one can write $r = r'r''$, $s = s's''$, where r'' and s'' have no prime factors in common and $r''s'' = m$. Now show that the element $\rho^{r'}\sigma^{s'}$ of G has order m. Deduce from this that G has an element whose order is divisible by the order of every element of G.

Apply this result in the case where G is a subgroup of the multiplicative group of a field F to show that G must then be a cyclic group and that, in fact, the elements of G are the roots in F of the polynomial $x^e - 1$, where e is the least common multiple of the orders of the elements of G.

Finally, deduce from this result that, for every positive integer n, there is a complex number γ such that $\gamma^n = 1$ and the powers $\gamma, \gamma^2, \ldots, \gamma^n$ are pairwise distinct (note that, in $\mathbf{C}[x]$, $x^n - 1$ shares no root with its derivative).

5. Suppose that L is a field, and that K is a subfield of L such that the dimension of L as a K-space is finite. Show that every element of L is a root of a polynomial with coefficients in K. Use this in order to prove that, if $K = \mathbf{R}$, then either $L = \mathbf{R}$, or L is isomorphic with $\mathbf{C}$.

6. Let F be a field, p a prime element of $F[x]$. Show that the factor ring $F[x]/(F[x]p)$ is a field, K say, containing a copy of F as a subfield, so that p may be regarded also as a polynomial with coefficients in K. With this understanding, show that p has a root in K.

Projects

1. Describe an integer computation accomplishing the following. For a polynomial function f with integer coefficients and a rational number of the form k/q, where k and q are integers and $q > 0$, the computation is to yield the signature of $f(k/q)$. Using this as a subroutine, design a computer program that locates the real roots of an element of $\mathbf{Q}[x]$ by the method of Section 2, with a guaranteed accuracy of $1/q$.

2. Consider the problem of finding a root of a polynomial function f on $\mathbf{C}$ with a computer program. The last part of the existence proof in Section 3 suggests the following procedure, in which one may use Exercise 1 above for calculating values of f and its derivatives. Start with an arbitrary complex number γ and calculate $f(\gamma)$. The coefficient of α^e in the formal expansion of $f(\gamma + \alpha)$ according to powers of α is equal to $\frac{1}{e!} f^{(e)}(\gamma)$, where $f^{(e)}$ denotes the eth derivative of f. This may be calculated by the recursion of Exercise 1, first for $e = 1$. If the result is found to be negligibly small, one calculates this for $e = 2$, etc. If k is the lowest index e for which the above expression, δ_k say, is not negligible, one uses an appropriate standard subroutine for calculating a complex number τ such that $\tau^k = -f(\gamma)/\delta_k$. Now one expects that, for a sufficiently small positive real number ρ (< 1), the complex number $\gamma + \rho\tau$ is closer to a root of f than γ. With due precautions in making the decisions alluded to above, iteration of this process eventually yields a good approximation to a root of f.

3. Suppose that points μ_i in $\mathbf{R}$ are given, such that $\alpha = \mu_1 < \cdots < \mu_n = \beta$, and corresponding values $f(\mu_1), \ldots, f(\mu_n)$ are given also. Consider the problem of finding a line L in $\mathbf{R}^2$ such that the maximum of the vertical distances of the points $(\mu_i, f(\mu_i))$ from L is as small as possible. First, complete the definition of the function f so that the graph of f in $\mathbf{R}^2$ is the union of the line segments from $(\mu_i, f(\mu_i))$ to $(\mu_{i+1}, f(\mu_{i+1}))$ for $i = 1, \ldots, n-1$. We assume that $n \geqq 3$, because the problem is trivial when $n < 3$. The required line L will be the graph of a polynomial function p of degree $\leqq 1$ on $\mathbf{R}$. Clearly p is a solution of our present problem if and only if it is the best possible approximation to f in the sense of Section 4. From there, we know that p is characterized by the existence of points γ_i such that $\alpha \leqq \gamma_0 < \gamma_1 < \gamma_2 \leqq \beta$ and $f(\gamma_i) - p(\gamma_i) = (-1)^i\sigma\|f - p\|$, where σ is either 1 or -1. From the geometry of our graphs here, it is clear that these points γ_i may be taken to be among the μ_j's. Hence, p is the solution if and only if there are indices $r < s < t$ such that

$$f(\mu_r) - p(\mu_r) = \sigma\|f - p\| = f(\mu_t) - p(\mu_t)$$

and

$$f(\mu_s) - p(\mu_s) = -\sigma\|f - p\|$$

Moreover, $\|f - p\|$ is simply the largest of the numbers $|f(\mu_i) - p(\mu_i)|$.

Now show that p is completely determined once the triple of indices (r, s, t) is known. Using this fact, make a computer program that simply selects triples (r, s, t) in turn, each time computes the values $p(\mu_i)$ for the corresponding p, and checks whether or not the above conditions are satisfied. We know that this must eventually find the solution p.

CHAPTER VII

The Exponential Function

1. Let z denote the identity map on $\mathbf{C}$. For every non-negative integer n, we define a polynomial function E_n by

$$E_n = \sum_{k=0}^{n} \frac{1}{k!} z^k$$

Given an arbitrary complex number c, let n be such that $n + 1 \geqq 2|c|$, and let q be an arbitrary positive integer. Then we have

$$\begin{aligned} |E_{n+q}(c) - E_n(c)| &\leqq \sum_{k=1}^{q} \frac{1}{(n+k)!} |c|^{n+k} \\ &\leqq \frac{|c|^{n+1}}{(n+1)!} \sum_{k=1}^{q} \left(\frac{|c|}{n+1}\right)^{k-1} \\ &\leqq 2 \frac{|c|^{n+1}}{(n+1)!} \end{aligned}$$

This shows that the sequence $(E_n(c))_{n=0,1,\ldots}$ is a Cauchy sequence of complex numbers. For every complex number c, we define the complex number $\exp(c)$ as the limit of this Cauchy sequence. The function exp from $\mathbf{C}$ to $\mathbf{C}$ so defined is the *exponential function*.

Now let u and v be complex numbers, and let us consider the product $E_n(u)E_n(v)$. It is easy to verify directly that, in the full expansion of this product according to the powers of u and v, the terms of total degree m in u and v add up to precisely $\frac{1}{m!}(u + v)^m$, as long as $m \leqq n$. Clearly, it follows from this that $\exp(u)\exp(v) = \exp(u + v)$. Thus, *the exponential function is a group homomorphism from the additive group of* $\mathbf{C}$ *to the multiplicative group of the non-zero elements of* **C** (since $\exp(0) = 1$, it is clear that $\exp(u) \neq 0$ for every complex number u).

From the above, we know that $|E_q(c) - E_0(c)| \leqq 2|c|$ for all positive integers q and all complex numbers c with $|c| \leqq 1/2$. It follows immediately from this that $|\exp(c) - 1| \leqq 2|c|$ whenever $|c| \leqq 1/2$. Since

$$\exp(u + c) - \exp(u) = \exp(u)(\exp(c) - 1)$$

for every u in **C**, this shows that *the exponential function is continuous.*

Remembering that the complex conjugation is a ring automorphism of **C**, one sees from the definition of exp that $\exp(c^*) = (\exp(c))^*$ for every complex number c. In particular, if c is of the form γi, with γ in **R**, we obtain

$$|\exp(\gamma i)|^2 = \exp(\gamma i)(\exp(\gamma i))^* = \exp(\gamma i)\exp(-\gamma i) = \exp(0) = 1.$$

Thus, *the map sending each real number* γ *to* $\exp(\gamma i)$ *is a continuous group homomorphism from the additive group of* **R** *to the multiplicative group of the complex numbers of absolute value* 1.

We wish to show that this group homomorphism is surjective. Write $\exp(\gamma i) = \cos(\gamma) + \sin(\gamma)i$, where $\cos(\gamma)$ and $\sin(\gamma)$ belong to **R**. This defines cos and sin as continuous real-valued functions on **R**. In order to facilitate estimation of values of cos and sin, we define two sequences of polynomial functions on **R**, as follows.

$$C_n = \sum_{k=0}^{n} \frac{(-1)^k}{(2k)!} t^{2k}$$

$$S_n = \sum_{k=0}^{n} \frac{(-1)^k}{(2k+1)!} t^{2k+1}$$

where t is the identity map on **R**. From our above discussion of the sequence E_n, we see that $(C_n(\gamma))$ and $(S_n(\gamma))$ are Cauchy sequences of real numbers for every real number γ, and that their limits are $\cos(\gamma)$ and $\sin(\gamma)$, respectively.

Note that if $|\gamma| \leqq 1$ then the terms in $C_n(\gamma)$ are non-increasing in absolute value and alternate in sign (unless $\gamma = 0$). In particular, it follows from this that $\cos(\gamma) \leqq C_2(\gamma)$ whenever $|\gamma| \leqq 1$. Using this, one verifies that

$$\cos(2/3) < \sqrt{3}/2.$$

Since cos is continuous, this implies that there is a real number γ (in $[0, 2/3]$) such that $\cos(\gamma) = \sqrt{3}/2$. The only complex numbers of absolute value 1 whose real parts are equal to $\sqrt{3}/2$ are $\exp(\gamma i)$ and $\exp(-\gamma i)$. Therefore, there is a real number μ such that $\exp(\mu i) = \sqrt{3}/2 + (1/2)i$. Now one verifies directly that the cube of this complex number is i, so that its sixth power is -1, which means that $\exp(6\mu i) = -1$. Using the continuity of cos as before, we conclude from this that, as γ ranges from 0 to 6μ, the real part of $\exp(\gamma i)$ takes on every value in $[-1, 1]$ at least once. The only complex numbers of absolute value 1 with real part $\cos(\gamma)$ are $\exp(\gamma i)$ and its reciprocal $\exp(-\gamma i)$.

We conclude that, for every complex number c of absolute value 1, there is a real number γ such that $\exp(\gamma i) = c$.

The real number π is defined as the greatest lower bound of the set of all positive real numbers ρ such that $\exp(\rho i) = -1$. It follows at once from the continuity of exp that $\exp(\pi i) = -1$, so that π is actually the smallest positive real number ρ with $\exp(\rho i) = -1$. It is easy to see from this that the kernel of the group homomorphism sending each real number γ to $\exp(\gamma i)$ consists precisely of the integer multiples of 2π.

On the other hand, let us consider the restriction of exp to $\mathbf{R}$. Evidently, as ρ increases from 0 without bound, $\exp(\rho)$ increases from 1 without bound. Moreover, we have $\exp(\rho) > 1$ whenever $\rho > 0$. Since exp is a group homomorphism, it is therefore strictly increasing on $\mathbf{R}$. Since $\exp(-\rho)$ is the reciprocal of $\exp(\rho)$, it follows that $\exp(-\rho)$ decreases from 1 as ρ increases from 0, and that the greatest lower bound of the set $\exp(\mathbf{R})$ is 0. Of course, every element of this set is positive. Every complex number c other than 0 can be written in exactly one way as a product ru, where r is a positive real number and $|u| = 1$. From this and the results established above, we have the following conclusions.

The function exp *is a surjective and continuous group homomorphism from the additive group of* $\mathbf{C}$ *to the multiplicative group of* $\mathbf{C}$. *The kernel of* exp *consists of the integer multiples of* $2\pi i$. *Finally,* $\exp(c)$ *is of absolute value* 1 *if and only if* c *is of the form* γi *with* γ *in* $\mathbf{R}$.

2. By a *simple continuous path* in $\mathbf{C}$ we mean an injective continuous map f from an interval $[\alpha, \beta]$ in $\mathbf{R}$ to $\mathbf{C}$. We claim that *the inverse map, g say, from $f([\alpha, \beta])$ to $[\alpha, \beta]$ is continuous.* Indeed, suppose that this is not the case. Then there is a positive real number ε and a point p in $f([\alpha, \beta])$ such that there are points q in $f([\alpha, \beta])$ arbitrarily close to p with $|g(p) - g(q)| > \varepsilon$. More specifically, for every positive integer n, there is a point q_n in $f([\alpha, \beta])$ such that $|g(p) - g(q_n)| > \varepsilon$ and $|p - q_n| < 1/n$. By the now familiar iterated bisection argument, we see that there is a point γ in $[\alpha, \beta]$ belonging to arbitrarily small intervals that contain points $g(q_n)$ for infinitely many n's. Now the continuity of f implies that $f(\gamma) = p$. On the other hand, it is clear that $|g(p) - \gamma| \geqq \varepsilon$, so that we have a contradiction to the injectivity of f. Our conclusion is that the inverse of f is continuous.

The above discussion provides the foundation for the notion of *betweenness* for points of $f([\alpha, \beta])$, which is actually independent of the particular choice of the map f. If p, q, r are points of $f([\alpha, \beta])$, we say that *q is between p and r if, for every continuous map h from an interval $[\gamma, \delta]$ of $\mathbf{R}$ to $f([\alpha, \beta])$ such that $h(\gamma) = p$ and $h(\delta) = r$, the image set $h([\gamma, \delta])$ also contains q.* We shall show that q is between p and r if and only if it belongs to $f([g(p), g(r)])$, where g still denotes the inverse of f.

By the very definition of between-ness, this condition is necessary. It remains to be shown that if this condition is satisfied, and h is as described in the definition, then q belongs to $h([\gamma, \delta])$. Consider the map $g \circ h$ from $[\gamma, \delta]$ to $[\alpha, \beta]$. Since this map is continuous, and since it takes the values $g(p)$ and $g(r)$ at the endpoints γ and δ of $[\gamma, \delta]$, it takes every value in $[g(p), g(r)]$ at some point of $[\gamma, \delta]$. By assumption, we have $q = f(\mu)$, with $\mu \in [g(p), g(r)]$. By what we have just shown, $\mu = g(h(\sigma))$, with $\sigma \in [\gamma, \delta]$. Hence, $q = f(g(h(\sigma))) = h(\sigma) \in h([\gamma, \delta])$, *q.e.d.*

Let S be the image of a simple continuous path in $\mathbf{C}$. A *progression of points* on S is a finite sequence $(s_0, \ldots, s_n)$ of points of S such that s_i is between s_{i-1} and s_{i+1} for every index i with $0 < i < n$, while s_0 and s_n are the extremities of S. The *length* of such a progression is defined as $\sum_{i=0}^{n-1} |s_{i+1} - s_i|$. The *mesh* of the progression is defined as the maximum of these n non-negative real numbers $|s_{i+1} - s_i|$.

At this point, we wish to use the fact that *a continuous function f on an interval $[\alpha, \beta]$ is uniformly continuous, in the sense that, for every positive real number ε, there is a positive real number δ such that $|f(\rho) - f(\sigma)| \leqq \varepsilon$ whenever $|\rho - \sigma| \leqq \delta$.* One proves this by assuming it to be false, and then using the argument of iterated interval bisection to show that this leads to the existence of a point in $[\alpha, \beta]$ at which the continuity property of f fails.

Returning to the above S, we see from the uniform continuity of a simple continuous path whose image is S that *there are progressions of arbitrarily small mesh on S.* Now suppose that there is a real number σ such that, for every positive real number ε, there is a positive real number δ such that the length of every progression of points on S whose mesh is less than δ differs from σ by less than ε. Then one says that S is *rectifiable*, and one calls σ the *length* of S.

The most important non-trivial example of a rectifiable path is a circular arc. Specifically, let α and β be real numbers, with $\alpha < \beta$ and $\beta - \alpha < 2\pi$. Define the map f from $[\alpha, \beta]$ to $\mathbf{C}$ by $f(\gamma) = \exp(\gamma i)$. We know from Section 1 that this is a simple continuous path in $\mathbf{C}$. Let us show that *this circular arc is rectifiable, and* that *its length is equal to $\beta - \alpha$.*

Consider a progression $\alpha = \gamma_0 \leqq \cdots \leqq \gamma_n = \beta$ of points on $[\alpha, \beta]$. Then $\exp(\gamma_0 i), \ldots, \exp(\gamma_n i)$ is a progression of points on our circular arc. We have

$$\exp(\gamma_{k+1} i) - \exp(\gamma_k i) = \exp(\gamma_k i)(\exp(\delta_k i) - 1)$$

where we have written δ_k for $\gamma_{k+1} - \gamma_k$. Directly from the definition of exp, we see that we may write $\exp(\delta_k i) - 1$ in the form $\delta_k i + \delta_k^2 M_k$, where M_k is a complex number with $|M_k| \leqq \exp(\delta_k) < \exp(2\pi) = M$, say. Hence, if δ is the maximum of the non-negative real numbers δ_k, we have

$$\delta_k(1 - \delta M) < |\exp(\gamma_{k+1} i) - \exp(\gamma_k i)| < \delta_k(1 + \delta M)$$

Clearly, every progression of points on our circular arc can be written as above, except for a possible inversion of its ordering. Using the continuity near the point 1 of the map sending each $\exp(\gamma i)$ to γ (for γ near 0 in $\mathbf{R}$), which is a special case of what we proved at the beginning of this section, we see that the above δ will be smaller than any arbitrarily fixed positive real number as soon as the mesh of the progression of points on our circular arc is sufficiently small. Since the length of the progression lies between $(\beta - \alpha)(1 - \delta M)$ and $(\beta - \alpha)(1 + \delta M)$, it follows that our arc is rectifiable and of length $\beta - \alpha$.

Now we have secured the foundation for the *numerical measure of angles*. Recall that an angle in $\mathbf{C}$ is an ordered pair (R, R') of rays with a common source s. Let r and r' be the points of R and R', respectively, whose distances from s are equal to 1. Then the numerical measure of the angle (R, R') is defined as the real number α such that $\exp(\alpha i) = (r - s)^{-1}(r' - s)$ and $0 \leqq \alpha < 2\pi$. By the above, α is equal to the length of the circular arc of radius 1, with s as center, drawn counterclockwise from R to R'.

Let T be a rotation of the plane. Then T associates an angle

$$(R_0(p), R_0(T(p)))$$

with each point p other than 0, where $R_0(q)$ denotes the ray with source 0 containing q. We know that T is the multiplication by a complex number t of absolute value 1. Substituting tp for $T(p)$ and using the above definition of numerical measure, we find that, if α is that numerical measure, then $\exp(\alpha i) = t$. Visually, this means that T is the counterclockwise turn through the angle of numerical measure α.

3. Eventually, we wish to discuss the (partial) inverse, log, of exp. For this, we shall use the following machinery of power series. Let R be an arbitrary ring, and recall that the ring of power series with coefficients in R is the ring $R[N^+]$ based on the monoid N^+. The generalities that follow are based on the discussion and notation of Section VI.1.

Let f and g be elements of $R[N^+]$, and assume that $g(0) = 0$. Then $g^k(n) = 0$ whenever $k > n$, so that a formally infinite linear combination of powers of g makes sense as an element of $R[N^+]$. With this understanding, we define the *composite* $[f, g]$ *of* f *with* g by the formula

$$[f, g](n) = \sum_{k \geqq 0} f(k) g^k(n)$$

Note that, if x has its usual meaning as an element of $R[N^+]$, then $[f, x] = f$.

From now on, we assume that R is commutative. Let δ be the canonical derivation of $R[N^+]$, so that $\delta(h)(n) = (n + 1)h(n + 1)$. We claim that

$$\delta([f, g]) = [\delta(f), g]\delta(g)$$

In verifying this, we shall write h' for $\delta(h)$. The definition gives

$$[f, g]'(n) = \sum_{k \geqq 0} f(k)(n + 1)g^k(n + 1) = \sum_{k \geqq 0} f(k)(g^k)'(n)$$

Slightly misusing our notation, we write this result simply as

$$[f, g]' = \sum_{k \geq 0} f(k)(g^k)'$$

Since δ is a derivation and $R[N^+]$ is now commutative, we have $(g^k)' = kg^{k-1}g'$. With this, we obtain

$$\begin{aligned}[f, g]' &= \sum_{k \geq 1} kf(k)g^{k-1}g' \\ &= \sum_{k \geq 1} f'(k-1)g^{k-1}g' = [f', g]g'\end{aligned}$$

so that our claim is established.

Now let F be any field containing the field $\mathbf{Q}$ of rational numbers. Working in $F[N^+]$, we consider the *exponential series*, i.e., the element Exp of $F[N^+]$ defined by

$$\operatorname{Exp}(n) = \frac{1}{n!}$$

The other power series of interest here will be denoted simply by L. It is defined by

$$L(0) = 0$$

$$L(n) = \frac{1}{n} \quad \text{for } n > 0$$

The clairvoyant reader will observe that, in the usual notation, we have $L = -\log(1 - x)$. Accordingly, we shall prove next that

$$[\operatorname{Exp}, -L] = 1 - x \quad \text{and} \quad [L, 1 - \operatorname{Exp}] = -x$$

First, note that $L'(n) = 1$ for every n. Hence, for every element f of $F[N^+]$, we have $(fL')(n) = \sum_{k=0}^{n} f(k)$. In particular, taking $f = 1 - x$, we find that $(1 - x)L' = 1$. On the other hand, the definition of Exp shows that $\operatorname{Exp}' = \operatorname{Exp}$.

Now let us write f for $[\operatorname{Exp}, -L]$. Then $f' = [\operatorname{Exp}', -L](-L)' = -fL'$, and hence $(1 - x)f' = -f$. Hence,

$$\begin{aligned}f(n) &= ((x - 1)f')(n) \\ &= \sum_{u+v=n} (x - 1)(u)f'(v)\end{aligned}$$

This yields $f(0) = -f'(0) = -f(1)$ and, for $n > 0$,

$$f(n) = -f'(n) + f'(n - 1) = -(n + 1)f(n + 1) + nf(n)$$

so that

$$(n - 1)f(n) = (n + 1)f(n + 1)$$

for every $n > 0$. This recursion shows that $f(n) = 0$ whenever $n > 1$. Directly from the definitions, one finds $f(0) = 1$. Hence, $f(1) = -1$, so that $f = 1 - x$.

Now let us write f for $[L, 1 - \mathrm{Exp}]$. Then we have

$$f' = [L', 1 - \mathrm{Exp}](1 - \mathrm{Exp})' = -[L', 1 - \mathrm{Exp}]\mathrm{Exp}$$

Next, we note that, for every element g of $F[N^+]$ with $g(0) = 0$, one has $[1 - x, g] = 1 - g$, as one sees at once from the definition of the composite and the fact that $(1 - x)(k) = 0$ whenever $k > 1$. In particular, we have $[1 - x, 1 - \mathrm{Exp}] = \mathrm{Exp}$, so that the above expression for f' yields

$$f' = -[L', 1 - \mathrm{Exp}][1 - x, 1 - \mathrm{Exp}]$$

It is easy to see from the definition of composites that, quite generally, $[f_1, g][f_2, g] = [f_1 f_2, g]$. Hence, we have $f' = -[L'(1 - x), 1 - \mathrm{Exp}]$. Since $L'(1 - x) = 1$, this yields $f' = -1$. Directly from the definition, we have $f(0) = 0$. Hence, we must have $f = -x$.

This completes the proof of the above two identities relating Exp and L via composition.

4. Now let us apply our power series results to the exponential function exp on **C**. In view of the power series L in $\mathbf{C}[N^+]$, we define a sequence of polynomial functions L_n on **C** by

$$L_n = \sum_{k=1}^{n} \frac{1}{k} z^k$$

When convenient, we interpret L_0 as 0. Let us fix a positive real number $\rho < 1$. Then, for every complex number c with $|c| \leqq \rho$ and every positive integer q, we have

$$\begin{aligned} |L_{n+q}(c) - L_n(c)| &\leqq \sum_{k=1}^{q} \frac{1}{n+k} |c|^{n+k} \\ &\leqq \frac{|c|^{n+1}}{n+1} \sum_{k=1}^{q} |c|^{k-1} \\ &\leqq \frac{|c|^{n+1}}{n+1} (1 - \rho)^{-1} \end{aligned}$$

This shows that the sequence with terms $L_n(c)$ is a Cauchy sequence of complex numbers for every complex number c with $|c| < 1$. We denote its limit by $-\log(1 - c)$. This defines the complex-valued function log on the subset of **C** consisting of the complex numbers v such that $|1 - v| < 1$. From the above estimate, we know that if $|c| \leqq \rho < 1$ then

$$|-\log(1 - c) - L_n(c)| \leqq \rho^{n+1}(n + 1)^{-1}(1 - \rho)^{-1}$$

for every non-negative integer n. Since every L_n is continuous, this shows that log is continuous.

At this point, we require a certain inequality concerning the coefficients of a polynomial function on $\mathbf{C}$. We consider an arbitrary such polynomial function $f = \sum_{q=0}^{n} \gamma_q z^q$. Let μ be a positive real number, and suppose that M_μ is an upper bound for the set of non-negative real numbers $|f(c)|$ with $|c| = \mu$. The result we wish to establish is that $|\gamma_q| \leqq M_\mu/\mu^q$ for each q.

Write c for $\exp\left(\dfrac{2\pi}{n+1} i\right)$. Then $\sum_{r=0}^{n} (c^e)^r$ is equal to $n + 1$ if e is an integer multiple of $n + 1$, because then each summand is equal to 1. On the other hand, if e is an integer not divisible by $n + 1$, then the above sum is equal to 0, as is seen by multiplying it with $c^e - 1$ and noting that this factor is not equal to 0.

Now we have

$$f(\mu c^r)c^{-kr} = \sum_{q=0}^{n} \gamma_q \mu^q (c^{q-k})^r$$

summing for $r = 0, \ldots, n$ and using the above remark, we find

$$(n + 1)\gamma_k \mu^k = \sum_{r=0}^{n} f(\mu c^r)c^{-kr}$$

Each term of the sum on the right is of absolute value $\leqq M_\mu$, so that this yields the inequality we wished to establish.

Now let us consider the composite polynomial function $f_n = E_n(-L_n)$ on the set of complex numbers c with $|c| \leqq \rho$, where ρ is a fixed positive real number <1. First, we show that these f_n's constitute a *Cauchy sequence of functions*, i.e., that they satisfy the following requirement. For every positive real number ε, there is an integer $N(\varepsilon)$ such that $|f_p(c) - f_q(c)| \leqq \varepsilon$ whenever p and q are no smaller than $N(\varepsilon)$ and $|c| \leqq \rho$.

In order to see this, we write $f_{n+q}(c) - f_n(c)$ in the form

$$[E_{n+q}(-L_{n+q}(c)) - E_n(-L_{n+q}(c))] + [E_n(-L_{n+q}(c)) - E_n(-L_n(c))]$$

We know from the above that $|L_m(c)| \leqq \rho/(1 - \rho)$ for every m, provided that $|c| \leqq \rho$. Therefore, the inequality at the beginning of Section 1 shows that the first of the two above differences is of absolute value no greater than

$$2(\rho/(1 - \rho))^{n+1}/(n + 1)!$$

In order to estimate the second difference above, we note that

$$E_n(u) - E_n(v) = \sum_{p=1}^{n} \frac{1}{p!}(u - v)\left(\sum_{r=1}^{p} u^{r-1}v^{p-r}\right)$$

It follows from this that the second difference above is of absolute value no greater than

$$|L_{n+q}(c) - L_n(c)| \sum_{p=1}^{n} \frac{1}{p!} p(\rho/(1-\rho))^{p-1} < |L_{n+q}(c) - L_n(c)| \exp\left(\frac{\rho}{\rho - 1}\right)$$

With our above estimate for the first factor here, we find that our difference is of absolute value smaller than

$$\rho^{n+1}(n+1)^{-1}(1-\rho)^{-1} \exp\left(\frac{\rho}{1-\rho}\right)$$

It is clear from these estimates that the f_n's constitute a Cauchy sequence of functions.

Now let $\gamma(n)_e$ denote the coefficient of z^e in the polynomial function f_n. Then, if p and q are indices no smaller than the above $N(\varepsilon)$, so that $|f_p(c) - f_q(c)| \leqq \varepsilon$ whenever $|c| \leqq \rho$, we know from what we have shown in general about the coefficients of polynomial functions that

$$|\gamma(p)_e - \gamma(q)_e| \leqq \varepsilon\rho^{-e}$$

for every non-negative index e.

From the formal identity $[\text{Exp}, -L] = 1 - x$, we see that, for each e, there is an index $\mathbf{Q}(e)$ such that, for every $q \geqq \mathbf{Q}(e)$, we have $\gamma(q)_e$ equal to 1, -1 or 0 according to whether e is 0, 1 or greater than 1. Consequently,

$$|\gamma(p)_0 - 1| \leqq \varepsilon, \qquad |\gamma(p)_1 + 1| \leqq \varepsilon\rho^{-1} \quad \text{and} \quad |\gamma(p)_e| \leqq \varepsilon\rho^{-e}$$

for each $e > 1$ whenever $p \geqq N(\varepsilon)$.

Now fix a positive real number $\sigma < \rho$. Our inequalities for the coefficients of f_p yield

$$|f_p(c) - (1 - c)| \leqq \varepsilon \sum_e (\sigma/\rho)^e < \varepsilon/(1 - (\sigma/\rho))$$

whenever $p \geqq N(\varepsilon)$ and $|c| \leqq \sigma$. Our conclusion is that, for every complex number c with $|c| < 1$, the sequence with terms $f_p(c)$ approaches the limit $1 - c$. On the other hand, it is clear from the definition of f_p that this sequence approaches the limit $\exp(\log(1 - c))$. Consequently, *we have*

$$\exp(\log(1 - c)) = 1 - c \quad \textit{whenever} \quad |c| < 1.$$

The composite log(exp) can be dealt with in almost the same way. Fix a positive real number $\eta < 1/2$. As we have already seen in Section 1, one has $|1 - \exp(c)| \leqq 2|c|$ whenever $|c| \leqq 1/2$. Here, we need only that

$$|1 - \exp(c)| \leqq 2\eta$$

whenever $|c| \leqq \eta$. It is easy to see from this that there is a positive real number $\tau < 1$ and a positive integer N, such that $|1 - E_m(c)| \leqq \tau$ whenever $m \geqq N$ and $|c| \leqq \eta$.

Now let us put $g_n = L_n(1 - E_n)$, and let us write $g_{n+q}(c) - g_n(c)$ in the form

$$[L_{n+q}(1 - E_{n+q}(c)) - L_n(1 - E_{n+q}(c))] + [L_n(1 - E_{n+q}(c)) - L_n(1 - E_n(c))]$$

For $n \geqq N$, and c such that $|c| \leqq \eta$, we find that the first of these differences is of absolute value no greater than $|c|^{n+1}(n + 1)^{-1}(1 - \tau)^{-1}$. In order to estimate the second of the above differences, we note first that

$$L_n(u) - L_n(v) = (u - v)\sum_{k=1}^{n} \frac{1}{k}\left(\sum_{r=1}^{k} u^{r-1}v^{k-r}\right)$$

Putting $u = 1 - E_{n+q}(c)$ and $v = 1 - E_n(v)$, and observing that then each of $|u|$ and $|v|$ is no greater than τ, we find that the absolute value of the second of the above differences is no greater than

$$|E_{n+q}(c) - E_n(c)|(1 - \tau)^{-1} \leqq 2|c|^{n+1}((n + 1)!)^{-1}(1 - \tau)^{-1}$$

Now it is clear that the sequence with terms g_n is a Cauchy sequence of polynomial functions on the set of complex numbers c with $|c| \leqq \eta$. Hence, we can proceed as we did with the sequence of f_n's and conclude from the identity $[L, 1 - \text{Exp}] = -x$ that *we have* $\log(\exp(c)) = c$ *whenever* $|c| < 1/2$.

The limitation on c we have made here is, of course, not the mildest possible one. However, note that $\log(\exp(c))$ makes sense whenever $|\exp(c) - 1| < 1$, while the equality cannot possibly hold subject to this restriction alone, because $\exp(c + 2\pi i) = \exp(c)$.

5. We wish to discuss the role played by polynomial and exponential functions in the context of differential equations. The formal part of this theory is quite analogous to the content of Section VI.5, where we treated polynomial and exponential sequences in the context of difference equations.

Let F be a field containing the field $\mathbf{Q}$ as a subfield. First, we work within $F[N^+]$. This contains the F-algebra $F[x]$ of polynomials as a sub F-algebra. For every element α of F, we denote by Exp_α the element of $F[N^+]$ defined by

$$\text{Exp}_\alpha(n) = \frac{\alpha^n}{n!}$$

where we agree that Exp_0 is the identity element of $F[N^+]$ (interpreting 0^0 as 1). Note that Exp_1 is the element Exp used before. It follows directly from the above definition that $\text{Exp}_\alpha \text{Exp}_\beta = \text{Exp}_{\alpha+\beta}$ for all elements α and β of F. The effect of the derivation δ is given by $\delta(\text{Exp}_\alpha) = \alpha \text{Exp}_\alpha$.

The analogue of the first result of Section VI.5 is that, *within* $F[N^+]$, *the set of elements* Exp_α *is linearly independent over* $F[x]$. The proof of this is so nearly the same as the proof in Section VI.5 that we need not write it out in detail. The place of the difference operator τ is here taken by the differential operator δ on $F[N^+]$. Since this acts as an ordinary derivation, the twist by the former shift operator σ is absent here. These are the only features in which the proof needed here differs from the proof of the analogue in Section VI.5.

Let I denote the identity map on $F[N^+]$, and consider the operator $\delta - \mu I$ on $F[N^+]$, where μ is an element of F. For every element p of $F[N^+]$, we have

$$(\delta - \mu I)(p \operatorname{Exp}_\mu) = \delta(p) \operatorname{Exp}_\mu$$

If p is a polynomial of degree d then $\delta(p)$ is a polynomial of degree $d - 1$, and it follows inductively that $(\delta - \mu I)^{d+1}$ annihilates $p \operatorname{Exp}_\mu$.

Now let us consider an operator T on $F[N^+]$ of the form

$$T = (\delta - \mu_1 I)^{e_1} \cdots (\delta - \mu_r I)^{e_r}$$

where the μ_i's are pairwise distinct elements of F and the e_i's are positive integers. It is clear from what we have just seen that T annihilates the elements $f_{i,s} = x^s \operatorname{Exp}_{\mu_i}$, where i ranges from 1 to r, and the index s associated with i ranges from 0 to $e_i - 1$. By the above, the set of these $f_{i,s}$'s is an F-linearly independent set, whose cardinality $\sum_{i=1}^r e_i$ equals the degree, n say, of T as a polynomial in δ.

Let $\mathbf{S}(T)$ denote the set of all elements f of $F[N^+]$ that are annihilated by T. Evidently, $\mathbf{S}(T)$ is a sub F-space of $F[N^+]$. If f is an element of $\mathbf{S}(T)$ then, for each exponent $e \geqq n$, the element $\delta^e(f)$ may be written as an F-linear combination of the elements $\delta^k(f)$ with $k < n$, in which the coefficients are independent of f. Consequently, an element of $\mathbf{S}(T)$ is completely determined by its n values at $0, \ldots, n - 1$. Therefore, the dimension of $\mathbf{S}(T)$ cannot exceed n. Since *the above* $f_{i,s}$*'s* are F-linearly independent, it is now clear that they *constitute an F-basis of* $\mathbf{S}(T)$.

From now on, we take our base field F to be the field $\mathbf{C}$ of complex numbers. An element f of $\mathbf{C}[N^+]$ associates a sequence $f^\sigma(c)$ in $\mathbf{C}$ with every complex number c, where

$$f^\sigma(c)_n = \sum_{k=0}^{n} f(k)c^k$$

We shall say that f is *analytic* if $f^\sigma(c)$ is a Cauchy sequence for every complex number c, and then we denote the limit of $f^\sigma(c)$ by $f^\circ(c)$. In this way, every analytic element f determines a complex-valued function f° on $\mathbf{C}$. If f belongs to $\mathbf{C}[x]$, then f° is, of course, the corresponding polynomial function on $\mathbf{C}$, as used before. Also, we have already seen before that the element Exp of $\mathbf{C}[N^+]$ is analytic, and we have $\operatorname{Exp}^\circ = \exp$. Hence, every $\operatorname{Exp}_\alpha$ is analytic, and we have $\operatorname{Exp}_\alpha(c) = \exp(\alpha c)$ for every complex number c.

Evidently, if f is analytic, then the sequence of non-negative real numbers $|f(n)|\rho^n$ has 0 as its limit for every non-negative real number ρ. An easy estimate of the kind we have made several times in Section 4 shows that this condition is also sufficient for f to be analytic. In particular, it follows almost immediately from this that sums and products of analytic elements are analytic, so that *the analytic elements constitute a sub* **C**-*algebra of* $\mathbf{C}[N^+]$. Moreover, one sees directly that *the map associating the function* f° *on* **C** *with every analytic element* f *is a homomorphism of* **C**-*algebras*. If f is an analytic element such that $f^\circ = 0$ then we see that we must have $f = 0$ by considering the sequence $f^\sigma(c)$ with c near 0. Thus, *the above homomorphism of* **C**-*algebras is injective*.

Next, it follows readily from our above criterion for analyticity that $\delta(f)$ is analytic whenever f is analytic. It is easy to verify that, in fact, $\delta(f)^\circ$ *is the derivative of* f°, *in the usual sense*. More precisely, for every complex number c, there are positive real numbers M_c and ε_c such that

$$|f^\circ(c+u) - f^\circ(c) - u\delta(f)^\circ(c)| \leqq |u|^2 M_c$$

whenever $|u| \leqq \varepsilon_c$.

Let us call the functions f° corresponding to the analytic elements f of $\mathbf{C}[N^+]$ the *power series functions* on **C**. Let D denote the differentiation operator on the **C**-algebra of all power series functions, so that $D(f^\circ) = \delta(f)^\circ$. In order to simplify notation, identify the elements of **C** with the corresponding scalar multiplications on the algebra of power series functions. Finally, write $\exp_\alpha$ for $\mathrm{Exp}_\alpha^\circ$. Then we have the following transcription of our above formal results.

Let $\mu_1, \ldots, \mu_r$ *be pairwise distinct complex numbers, and let* $e_1, \ldots, e_r$ *be positive integers. Then the power series functions annihilated by the operator* $(D - \mu_1)^{e_1} \cdots (D - \mu_r)^{e_r}$ *constitute a* **C**-*space with* **C**-*basis* $z^{s_i} \exp_{\mu_i}$, *where* $0 \leqq s_i < e_i$.

Exercises

1. Directly from the definitions, show that

$$\cos((n+1)\alpha) + \cos((n-1)\alpha) = 2\cos(n\alpha)\cos(\alpha)$$

for every integer n and every real number α. Deduce that, for every nonnegative integer n, there is a polynomial function T_n on **R**, of degree n and with integer coefficients, such that

$$T_n(\cos(\alpha)) = \cos(n\alpha)$$

for every real number α. More precisely, if t denotes the identity map on **R**, show that the T_n's are given recursively by

$$T_0 = 1, \qquad T_1 = t$$

$$T_{n+2} = 2tT_{n+1} - T_n$$

These polynomial functions are known as the *Chebyshev Polynomials*. An indication of their significance is contained in the next exercise.

2. For T_n as in Exercise 1, with $n > 0$, show that, in $[-1, 1]$, the maximum absolute value of T_n is equal to 1 and is attained at the $n + 1$ points $\eta_{0,n}$, $\eta_{1,n}, \ldots, \eta_{n,n}$, where

$$\eta_{k,n} = \cos\left(\frac{k}{n}\pi\right)$$

with $T_n(\eta_{k,n}) = (-1)^k$. Deduce that if p is any polynomial function of degree $n > 0$, with highest coefficient 1 and all coefficients in **R**, then the maximum absolute value of p on $[-1, 1]$ is no smaller than 2^{1-n}. [Consider the polynomial function $f = p - 2^{1-n}T_n$, noting that its degree is strictly smaller than n. If the claim for p were false, f would have at least n zeros.]

3. Directly from the result of Section 2 concerning the length of a circular arc, show that $\alpha \leqq 1 - \cos(\alpha) + \sin(\alpha)$ for every α in $[0, \pi/2]$.

4. Let f be a continuous group endomorphism of the additive group of **R**. By considering the restriction of f to **Q**, show that there is a real number μ such that $f(\rho) = \mu\rho$ for every real number ρ.

Now recall from Section 1 that the restriction of exp to **R** is a group isomorphism from the additive group of **R** to the multiplicative group, P say, of all positive real numbers. Denote the inverse homomorphism from P to **R** by ln. Show that, for every real number ρ with $|1 - \rho| < 1$, one has $\ln(\rho) = \log(\rho)$, and deduce that ln is a continuous function. Finally, apply the first part of this exercise to show that, if g is any continuous group homomorphism from **R** to P, there is a real number μ such that $g(\rho) = \exp(\mu\rho)$ for every real number ρ.

5. Let F stand for either of the fields **R** or **C**, and let A be a finite-dimensional F-algebra. Let α be an F-algebra derivation of A, i.e., an element of $\mathrm{End}_F(A)$ satisfying $\alpha(xy) = \alpha(x)y + x\alpha(y)$ for all elements x and y of A. Show that $\sum_{n \geqq 0} \frac{1}{n!}\alpha^n$ makes sense as an element $\exp(\alpha)$ of $\mathrm{End}_F(A)$, and that $\exp(\alpha)$ is actually an automorphism of F-algebras.

6. Let F be as in Exercise 5, and let V be an F-space of finite dimension d. Let α be a linear endomorphism of V, and note that it follows as in Exercise 5 that $\sum_{n \geqq 0} \frac{1}{n!}\alpha^n$ makes sense as a linear automorphism of V. Let α' denote the F-algebra derivation of the exterior algebra $E(V)$ that coincides with α on V. Show that $\exp(\alpha') = E(\exp(\alpha))$, where the expression on the right is defined from $\exp(\alpha)$ as in Section V.2. Next, observe that α' acts on the 1-dimensional space $E^d(V)$ as the scalar multiplication by a certain element $T(\alpha)$ of F (which is called the *trace* of α), and deduce from the last result that the determinant of $\exp(\alpha)$ equals $\exp(T(\alpha))$.

Projects

1. Suppose that computer routines for evaluating sin and cos are available. Examine the following program. Given real numbers α and β as inputs, where α lies in $[0, 1]$ and $\alpha^2 + \beta^2 = 1$, the program calculates real numbers γ_n by the following recursion: $\gamma_0 = \beta$, $\gamma_{n+1} = \gamma_n + \beta\cos(\gamma_n) - \alpha\sin(\gamma_n)$. In particular, verify that the sequence with terms γ_n approaches a real number γ in $[-\pi/2, \pi/2]$ such that $\cos(\gamma) = \alpha$ and $\sin(\gamma) = \beta$. Augment the program by a testing routine which results in termination of the program as soon as γ_n differs from the correct γ by no more than a prescribed allowable error.

2. Let $\eta_n = \exp(-(2\pi/n)i)$, with $n > 0$. Suppose that f is a complex-valued function on $\mathbf{R}$ that is a polynomial of degree less than n in the function e, where $e(\rho) = \exp(2\pi\rho i)$ for every real number ρ. We have seen in Section 4 (using a different notation) that the coefficient of e^p in f is equal to $(1/n)\sum_{r=0}^{n-1} f(r/n)(\eta_n)^{pr}$. This brings up the following computational task. Given are the powers $(\eta_n)^p$ and complex numbers a_p, for $p = 0, \ldots, n-1$. Required are the n complex numbers $c_p = \sum_{r=0}^{n-1} a_r(\eta_n)^{pr}$.

The usual calculation of each c_p is based on the recursion $u_0 = a_{n-1}$, $u_{q+1} = u_q(\eta_n)^p + a_{n-q-2}$, which yields c_p as u_{n-1}. If a *step* is understood to consist in carrying out one multiplication, followed by one addition, then the calculation of each c_p requires $n - 1$ steps, so that the total number of steps required in $n(n-1)$.

Now suppose that $n = st$, where s and t are positive integers. Verify that

$$c_p = \sum_{h=0}^{s-1}\left[\sum_{k=0}^{t-1} a_{ks+h}(\eta_t)^{p'k}\right](\eta_n)^{ph}$$

where p' is the non-negative remainder in the division of p by t. Deduce from this that the calculation of the c_p's can be so organized that the total number of steps required is only $n(s + t - 2)$. This reduction may be iterated, notably in the case where $n = 2^m$, in which case the calculation can be carried out in no more than nm steps. Implement this reduction with a recursive program design.

CHAPTER VIII

Integration

1. Let S be a subset of a metric space V. A point p of S is called an *interior point* of S if there is a positive real number δ such that every point of V whose distance from p is less than δ belongs to S. The set S is said to be *open* in V if every point of S is an interior point of S. If p is a point in V then every subset S of V containing p as an interior point is called a *neighborhood* of p in V. A subset of V is said to be *closed* in V if its complement in V is open.

Let $(\alpha_1, \ldots, \alpha_n)$ be a point of $\mathbf{R}^n$, and let $\sigma_1, \ldots, \sigma_n$ be positive real numbers. If B is the set of points $(\beta_1, \ldots, \beta_n)$ such that $|\beta_i - \alpha_i| \leqq \sigma_i$ for each index i then we call B a *block* in $\mathbf{R}^n$, and we refer to $(\alpha_1, \ldots, \alpha_n)$ as the *center* of B. Evidently, B is closed in $\mathbf{R}^n$, and the interior points of B are the points $(\beta_1, \ldots, \beta_n)$ such that $|\beta_i - \alpha_i| < \sigma_i$ for each i. By the definition of *volume* in $\mathbf{R}^n$, the volume of B is equal to the product of the lengths of the edges issuing from one vertex of B, so that the volume of B is equal to $2^n\sigma_1 \cdots \sigma_n$.

We shall be concerned only with the most direct extension of the notion of volume, which is as follows. Let S be a bounded subset of $\mathbf{R}^n$, so that S is contained in some block, B say. Consider finite families $\mathbf{K}$ of blocks whose interiors are pairwise disjoint and contained in S. The union of (the members of) $\mathbf{K}$ is a subset of $\mathbf{R}^n$ whose volume, by any reasonable definition, is equal to the sum of the volumes of the members of $\mathbf{K}$. It cannot exceed the volume of B. Therefore, the set of these volumes, for all families $\mathbf{K}$ as described above, has a least upper bound. This is called the *inner measure* of S, and we denote it by $m(S)$.

On the other hand, consider finite families $\mathbf{L}$ of blocks such that every point of S belongs to some member of $\mathbf{L}$, which circumstance we express by saying that the union of $\mathbf{L}$ contains S. Let $M(S)$ be the greatest lower bound of the set of non-negative real numbers obtained by adding the volumes of the members of $\mathbf{L}$, for each possible $\mathbf{L}$. This real number $M(S)$ is called the *outer measure* of S. It is not difficult to see that $m(S) \leqq M(S)$. Finally, if S is such that $m(S) = M(S)$, then one says that S *has volume* (relative to $\mathbf{R}^n$, or n-volume) and one calls the non-negative real number $m(S)$ the *volume* of S.

The property of having volume, for a bounded set S, may be elucidated as follows. Say that a point p of $\mathbf{R}^n$ belongs to the *boundary* of S if every neighborhood of p meets both S and the complement of S. Then, with some care, one can see that S has volume if and only if, for every positive real number ε, there is a finite family $\mathbf{F}_\varepsilon$ of blocks such that the union of $\mathbf{F}_\varepsilon$ contains the boundary of S, and the sum of the volumes of the members of $\mathbf{F}_\varepsilon$ does not exceed ε.

2. A simple but non-trivial example of a set having volume is as follows. Suppose that f is a continuous real-valued function on $\mathbf{R}^n$ satisfying the following conditions: (1) $f(0) = 0$ and $f(p) > 0$ whenever $p \neq 0$; (2) for every non-negative real number ρ and every point p, one has $f(\rho p) = \rho f(p)$. We shall show that the set S of all points p of $\mathbf{R}^n$ such that $f(p) \leqq 1$ has volume.

First, we make some elementary observations concerning unions of finite families of blocks in $\mathbf{R}^n$. Let $\mathbf{F}$ be any such family. Draw all the coordinate hyperplanes in $\mathbf{R}^n$ that contain a face of a member of $\mathbf{F}$. This yields a partitioning of each member of $\mathbf{F}$ into a union of subblocks, such that every intersection of members of $\mathbf{F}$ is a union of a family of these subblocks. Thus, the totality of subblocks so obtained is a new finite family of blocks *with pairwise disjoint interiors* whose union coincides with the union of $\mathbf{F}$. Next, we can cut up each member of our new family into subblocks of diameter not exceeding any preassigned positive real number, τ say. Let us call a family having these properties a *regular τ-family of blocks*. Now our remarks may be summarized as follows. *For every finite family* $\mathbf{F}$ *of blocks, and every positive real number τ, there is a regular τ-family of blocks whose union coincides with that of* $\mathbf{F}$ *and whose volume is no greater than the sum of the volumes of the members of* $\mathbf{F}$. Note that this shows that the union of $\mathbf{F}$ has volume, and that its volume is equal to the sum of the volumes of the members of the regular τ-family.

For every point p of $\mathbf{R}^n$, we denote the distance of p from 0 by $|p|$. From the continuity of f and the fact that $f(0) = 0$, it is clear that there is a positive real number σ such that every point p with $|p| \leqq \sigma$ belongs to S.

In order to proceed, we require a little more analysis, as follows. Clearly, every Cauchy sequence in $\mathbf{R}^n$ has a limit in $\mathbf{R}^n$. If T is a closed subset of $\mathbf{R}^n$ then the limit of every Cauchy sequence of points of T evidently belongs to T. Now suppose that T is bounded, as well as closed. Then one shows, as in the case of a closed interval in the place of T, that *if g is a continuous function on T then g is uniformly continuous* (cf. Section VII.2), *and g attains a minimum, as well as a maximum on T.*

In particular, the continuous function f attains a minimum on the set of points p with $|p| = 1$, and it is clear that this minimum must be a *positive* real number, μ say. Now it follows from condition (2) above that the points of S are at distances no greater than μ^{-1} from 0. Thus, S is bounded. Let C be a cubical block with 0 as center such that $(1/2)C$ contains S. Let σ be half of the edge length of C. Given an arbitrary positive real number α, we appeal to the uniform continuity of f on C and choose a positive real number $\tau \leqq \sigma/2$ such that $|f(p) - f(q)| \leqq \alpha$ whenever p and q are in C and $|p - q| \leqq \tau$.

Now consider a regular τ-family **F** of blocks such that the union of **F** contains S and every member of **F** contains a point of S. Let B be a member of **F**, and let p be a point of $B \cap S$. Let q be any point of B. Then we have $|p - q| \leqq \tau$. Since $\tau \leqq \sigma/2$, this implies that q belongs to C, and hence that $f(q) \leqq f(p) + \alpha \leqq 1 + \alpha$. Using condition (2), we deduce from this that $f((1 + \alpha)^{-1}q) \leqq 1$, i.e., that $(1 + \alpha)^{-1}q$ belongs to S. Our conclusion is that $(1 + \alpha)^{-1} B$ is contained in S.

Now let L denote the sum of the volumes of the members of **F**. Then the sum of the volumes of the blocks $(1 + \alpha)^{-1}B$ with B in **F** is evidently equal to $(1 + \alpha)^{-n}L$. Since all these blocks lie in S and since their interiors are pairwise disjoint, we have $(1 + \alpha)^{-n}L \leqq m(S)$. On the other hand, we have $L \geqq M(S)$. Consequently, $M(S) \leqq (1 + \alpha)^n m(S)$. We know from the definitions that $m(S) \leqq M(S)$. Since α may be taken arbitrarily small, it follows therefore from what we have just shown that $m(S) = M(S)$. The existence of regular τ-families **F** as used above being assured by the observations at the outset, this proves that S has volume.

3. A subset T of an **R**-space is called *convex* if it has the property that every line segment whose extremities belong to T is entirely contained in T. We establish a number of geometrical properties of convex sets, eventually obtaining the result that *every closed, bounded and convex subset of* $\mathbf{R}^n$ *has volume*.

Suppose that T is a non-empty open convex subset of $\mathbf{R}^n$ not containing 0. We claim that *there is a linear real-valued function* f *on* $\mathbf{R}^n$ *such that* $f(p) > 0$ *for every point* p *of* T. In order to see this, choose a sub **R**-space V of $\mathbf{R}^n$ such that $V \cap T = \varnothing$ and V has the largest possible dimension. Let η denote the canonical homomorphism of **R**-spaces from $\mathbf{R}^n$ to $\mathbf{R}^n/V$. Evidently, $\eta(T)$ is a non-empty convex subset of $\mathbf{R}^n/V$ not containing the 0-element V of $\mathbf{R}^n/V$. Moreover, if we regard $\mathbf{R}^n/V$ as an $\mathbf{R}^m$ by choosing an **R**-basis, then $\eta(T)$ is open in $\mathbf{R}^n/V$. Since a linear function on $\mathbf{R}^n/V$ that is positive on $\eta(T)$ yields a linear function on $\mathbf{R}^n$ that is positive on T, by composition with η, we have now reduced the situation to the case where every 1-dimensional sub **R**-space of the containing **R**-space meets the given convex subset. Finally, we replace T with the set T' consisting of all points αt with t in T and α a positive real number. It is easy to see that T' is still open and convex.

Evidently it does not contain 0. A linear function f satisfies the above requirements with respect to T if and only if it satisfies these requirements with respect to T'.

Accordingly, we suppose now that, if a point p belongs to T, so does every αp with $\alpha > 0$, and that every 1-dimensional subspace of $\mathbf{R}^n$ meets T. Under these assumptions, we shall show that we must have $n = 1$.

The usual iterated bisection argument shows that every line segment joining a point of T to a point not in T must contain a point of the boundary of T. Choose some point p of T. Since T is convex and does not contain 0, the point $-p$ does not belong to T. Now suppose that $n > 1$. Then there is a point q in $\mathbf{R}^n$ that is not a scalar multiple of p. If q does not belong to T, the line segment $[p, q]$ must contain a point of the boundary of T. If q belongs to T then, because $-p$ does not belong to T, the line segment $[q, -p]$ contains a point of the boundary of T. Thus, in either case, there is a point, r say, in $\mathbf{R}^n$ such that $r \neq 0$ and r belongs to the boundary of T. By one of our special assumptions, T contains a scalar multiple αr of r. Since T is open, the boundary point r cannot belong to T. Now it follows from our other special assumption that α must be negative, and hence that $-r$ belongs to T. Since T is open, there is a neighborhood, W say, of 0 in $\mathbf{R}^n$ such that $-r + W \subset T$. Now $r - W$ is a neighborhood of r, and therefore contains a point s of T. But $-s$ lies in $-r + W$ and thus belongs to T. By the convexity of T, this implies the contradiction that 0 belongs to T. Our conclusion is that $n = 1$.

Now we have $T \subset \mathbf{R}$. Since T is convex and does not contain 0, all the points of T have the same sign. Thus, either the identity map on $\mathbf{R}$ or else its negative is a linear function f with the required property.

Now let C be a closed convex subset of $\mathbf{R}^n$, and let C° denote the set of interior points of C. We assume that C° is non-empty, and we fix a point p of C°. Let q be a point of C. We claim that every point of the line segment $[p, q]$, with the possible exception of q, belongs to C°. In order to see this, consider a point $r = (1 - \alpha)p + \alpha q$, where $0 \leqq \alpha < 1$. Choose a neighborhood W of p such that $W \subset C$. Define the map g from $\mathbf{R}^n$ to $\mathbf{R}^n$ by $g(s) = (1 - \alpha)s + \alpha q$. Then $g(p) = r$ and, for every point s of $\mathbf{R}^n$, the image $g(s)$ belongs to the line segment $[s, q]$. Consequently, $g(C) \subset C$. Since $1 - \alpha \neq 0$ and W is a neighborhood of p, it is clear from the definition of g that $g(W)$ is a neighborhood of r. With $g(W) \subset C$, this shows that r belongs to C°, so that our above claim is now established.

In particular, what we have just proved implies that C° is convex, and that the points of C that do not belong to C° belong to the boundary of C°. From the fact that C is closed, it is clear that the boundary of C is contained in C. A fortiori, the boundary of C° is contained in C.

Suppose that q is a point of the boundary of C, and consider the subset $-q + C^\circ$ of $\mathbf{R}^n$. This is open and convex, and it does not contain 0. From the beginning of this section, we know that there is a real-valued linear function f on $\mathbf{R}^n$ such that $f(r) > 0$ for every point r in $-q + C^\circ$. Now let

p be a point of C°, and consider the ray with source p that contains q. The points of this ray may be written in the form

$$r(\alpha) = (1 - \alpha)p + \alpha q$$

where α ranges over the set of all non-negative real numbers. By the above, $r(\alpha)$ belongs to C° whenever $0 \leqq \alpha < 1$. On the other hand, we have $-q + r(\alpha) = (1 - \alpha)(-q + p)$, so that $f(-q + r(\alpha)) = (1 - \alpha)f(-q + p)$. This shows that, if $\alpha > 1$, then the value of f at $-q + r(\alpha)$ is negative. Since f is continuous and since its values on C° are positive, it follows that, for $\alpha > 1$, the point $r(\alpha)$ cannot belong to either C° or the boundary of C°, i.e., $r(\alpha)$ does not belong to C.

Now let us suppose that our closed convex set C is bounded and contains 0 as an interior point. For every point q of $\mathbf{R}^n$ other than 0, the ray with source 0 that contains q contains exactly one point, q^* say, of the boundary of C. Indeed, the boundedness of C implies that such a ray contains at least one point of the boundary of C, and we have just proved that a ray whose source is an interior point of C and which contains a point of the boundary of C contains only one point of the boundary of C. We define the real-valued function f on $\mathbf{R}^n$ by setting $f(0) = 0$ and $f(q) = |q|/|q^*|$ for every q in $\mathbf{R}^n$ other than 0. We shall prove that f is continuous.

From the fact that 0 is an interior point of C, it is easy to deduce that f is continuous at 0. Therefore, the continuity of f will be clear once we have shown that, for all points p and q of $\mathbf{R}^n$, we have $f(p + q) \leqq f(p) + f(q)$. It is clear from the definition that $f(r) > 0$ for every r other than 0, and that $f(\rho r) = \rho f(r)$ whenever ρ is a non-negative real number. Now suppose that q is a scalar multiple αp of p, where $1 + \alpha \geqq 0$. Then we have

$$f(p + q) = (1 + \alpha)f(p) = f(p) + \alpha f(p)$$

If $\alpha \geqq 0$ the last expression is equal to $f(p) + f(q)$. If $\alpha < 0$ this expression is no greater than $f(p)$ and hence no greater than $f(p) + f(q)$. Next, observe that if $q = \alpha p$ and $1 + \alpha < 0$ we have $p = \alpha^{-1}q$ and $1 + \alpha^{-1} > 0$. By the symmetry with respect to p and q, we conclude that now it suffices to prove the above inequality in the case where p and q are linearly independent.

Assume this to be the case. Then $p + q \neq 0$, and we know from the above that there is a linear function h on $\mathbf{R}^n$ such that $h(r) > h((p + q)^*)$ for every point r of C°. Since h is linear and the zero point belongs to C°, we must have $h((p + q)^*) < 0$. Now, from the equality

$$p + q = f(p + q)(p + q)^*$$

with $f(p + q) > 0$, it follows that $h(p + q) < 0$. This implies that at least one of $h(p)$ or $h(q)$ is negative. Interchanging p and q, if necessary, we arrange to have $h(q) < 0$. Let ρ be the positive real number such that $\rho h(p + q) = h(q)$.

Now let us first consider the case where $h(p) \geqq 0$. Then $h(p + q) \geqq h(q)$, so that $\rho \geqq 1$. Writing t for $\rho(p + q)$, we have $t^* = (p + q)^*$. Since every neighborhood of q^* contains a point of C°, we must have $h(q^*) \geqq h((p+q)^*)$, i.e., $h(q^*) \geqq h(t^*)$, whence $h(q^*)/h(t^*) \leqq 1$. On the other hand, $q = f(q)q^*$ and $t = f(t)t^*$, so that $f(q)h(q^*) = f(t)h(t^*)$. With the last inequality above, this yields $f(t) \leqq f(q)$, i.e., $\rho f(p+q) \leqq f(q)$. Since $\rho \geqq 1$, this gives $f(p+q) \leqq f(q)$.

It remains to deal with the case where $h(q) < 0$ and $h(p) < 0$. In this case, there are positive real numbers α and β such that $\alpha h(p) = h((p+q)^*) = \beta h(q)$. Since each of $h(p^*)$ and $h(q^*)$ is greater than or equal to $h((p+q)^*)$, it follows that we must have $\alpha f(p) \geqq 1$ and $\beta f(q) \geqq 1$. Now we have, from the above defining equalities for α and β,

$$h(p+q) = (\alpha^{-1} + \beta^{-1})h((p+q)^*)$$

Since $h(p+q) = f(p+q)h((p+q)^*)$, this yields

$$f(p+q) = \alpha^{-1} + \beta^{-1} \leqq f(p) + f(q)$$

This completes the proof of the general inequality for f. It follows from this that, for all points p and q of $\mathbf{R}^n$,

$$f(p) - f(-q) \leqq f(p+q) \leqq f(p) + f(q)$$

Together with the continuity of f at 0, this shows that f is continuous at p.

Clearly, our closed, bounded and convex set C is the set of all points p in $\mathbf{R}^n$ such that $f(p) \leqq 1$. We have shown that f satisfies all the conditions we made on the function dealt with in Section 2. Therefore, we have the conclusion that C has volume. This establishes the result announced at the beginning of this section. One must merely add the remark that, if the given set C has no interior point, then it is contained in some hyperplane of $\mathbf{R}^n$ and consequently has volume 0.

4. Let D be a closed and bounded set in $\mathbf{R}^n$ having volume, and suppose that f is a continuous real-valued function on D such that $f(p) \geqq 0$ for every point of D. We consider the subset $C(f)$ of $\mathbf{R}^{n+1} = \mathbf{R}^n \times \mathbf{R}$ whose points are the pairs (p, ρ), where p belongs to D and $0 \leqq \rho \leqq f(p)$. We wish to show that $C(f)$ has volume.

Let ε be a positive real number. We know that f is uniformly continuous on D. Accordingly, we consider positive real numbers τ such that $|f(p) - f(q)| \leqq \varepsilon$ whenever $|p - q| \leqq \tau$. Referring to Section 2, let $\mathbf{F}$ be a regular τ-family of blocks in $\mathbf{R}^n$ whose union contains D and every member of which meets D. For each member B of $\mathbf{F}$, let m_B and M_B be the minimum and the maximum of f on $B \cap D$. Let B^+ denote the block $B \times [0, M_B]$ in $\mathbf{R}^{n+1}$, and let B^- denote the block $B \times [0, m_B]$. Let Σ^+ denote the sum of the volumes of all the boxes B^+, and let Σ^- denote the sum of the volumes of the boxes B^- for which the interior of B is contained in D. Let $V(D)$ denote the n-volume of D. We know that τ and $\mathbf{F}$ can be so chosen that the sum of the volumes of all the boxes B is no greater than $V(D) + \varepsilon$, and the sum of the volumes of the boxes B whose interiors are contained in D is no smaller than $V(D) - \varepsilon$. Since each $M_B - m_B$ is no greater than ε, it follows then that we have

$$\Sigma^+ - \Sigma^- \leqq \varepsilon(V(D) + \varepsilon) + 2\varepsilon M$$

where M is the maximum of f on D. Evidently, the outer measure of $C(f)$ is no greater than Σ^+, while the inner measure of $C(f)$ is no smaller than Σ^-. Since ε is an arbitrary positive number, it follows that $C(f)$ has volume.

We denote the volume of $C(f)$ by $I_D(f)$ and call it the *integral* of f over D. This definition is extended to arbitrary continuous functions from D to $\mathbf{R}$, as follows. If f is such a function, define the function $|f|$ by $|f|(p) = |f(p)|$. Evidently, both $|f|$ and $f + |f|$ are continuous non-negative functions on D. We set

$$I_D(f) = I_D(f + |f|) - I_D(|f|)$$

This defines I_D as an $\mathbf{R}$-linear real-valued function on the $\mathbf{R}$-space of all continuous real-valued functions on D. The required verification can be carried out quite directly, using the fact that, if g is *any* non-negative continuous function on D such that $f + g$ is non-negative, then $I_D(f) = I_D(f + g) - I_D(g)$. Note also that the definition yields $I_D(f) \leqq I_D(|f|)$. Replacing f with $-f$, we obtain $-I_D(f) \leqq I_D(|f|)$. Hence, $|I_D(f)| \leqq I_D(|f|)$.

In order to systematize the calculation of volumes, one requires a little more information concerning the use of families of blocks. We obtain this from a general result which is actually of independent interest. This result is known as Lebesgue's Lemma and says the following.

Suppose that A is a closed and bounded subset of $\mathbf{R}^n$, and that $\mathbf{G}$ is a family of open subsets of $\mathbf{R}^n$ whose union contains A. There is a positive real number δ such that every subset of $\mathbf{R}^n$ having diameter no greater than δ and meeting A is contained in a member of $\mathbf{G}$.

We prove this result by deriving a contradiction from the assumption that it is false. Under this assumption, there is, for every positive integer q, a point a_q in A with the property that, for every member G of $\mathbf{G}$, there is a point in $\mathbf{R}^n$ not belonging to G and being at a distance no greater than $1/q$ from a_q. Since A is closed and bounded, we can apply the now familiar iterated partition argument to conclude that there is a point p in A such that every neighborhood of p contains a_q for infinitely many indices q. This implies that, for every member G of $\mathbf{G}$, every neighborhood of p contains a point not belonging to G. In other words, no neighborhood of p is contained in a member of $\mathbf{G}$, which contradicts the assumptions made concerning $\mathbf{G}$ in the statement of the result.

Now let $\mathbf{L}$ be a finite family of blocks whose union contains A. Let σ be a positive real number. For each member L of $\mathbf{L}$, let L^σ denote the block whose center coincides with the center of L and whose edges are of the lengths obtained by multiplying the lengths of the corresponding edges of L by $1 + \sigma$. Let $\mathbf{L}^\sigma$ denote the family of these new blocks L^σ, and let $\mathbf{G}$ be the family of the interiors of the members of $\mathbf{L}^\sigma$. This satisfies the requirements for the result we have just proved. Let δ be the bound obtained in that result, and let $\mathbf{F}$ be a regular δ-family of blocks (in the sense of Section 2) whose union contains A and every member of which has a point in common with A. Then every member of $\mathbf{F}$ is contained in some member of $\mathbf{G}$, and thus is contained in one of the L^σ's. Consequently, the volume of the union of $\mathbf{F}$ is no greater than the sum of the volumes of the blocks L^σ. This sum is equal to the product of $(1 + \sigma)^n$ and the sum of the volumes of the blocks L. Since σ is an arbitrary positive real number, we can draw the following general conclusion.

Suppose that A is a closed bounded subset of $\mathbf{R}^n$. For every positive real number ε, there is a positive real number τ such that, for every regular τ-family $\mathbf{F}$ of blocks whose union contains A and every member of which meets A, the volume of the union of $\mathbf{F}$ differs from the outer measure of A by no more than ε.

Now let us assume, in addition, that A has volume. Given a positive real number ε, let $\mathbf{G}$ be a regular family of blocks contained in A such that the volume of the union of $\mathbf{G}$ is no smaller than $V(A) - \varepsilon$, where $V(A)$ denotes the volume of A. Let σ be a real number such that $0 < \sigma < 1$. For every member G of $\mathbf{G}$, let G_σ denote the block whose center coincides with that of G and whose edges are of the lengths obtained by multiplying the lengths of the corresponding edges of G by σ. Let $\mathbf{G}_\sigma$ denote the family of all these new blocks G_σ. Now the union of $\mathbf{G}_\sigma$ is a closed set in $\mathbf{R}^n$ not meeting the boundary of A. The function, D say, whose value at each point p of the boundary of A is the distance from p to the union of $\mathbf{G}_\sigma$ is continuous, and all its values are positive. Since the boundary of A is closed and bounded, it follows that this function D has a positive minimum on the boundary of A. Let δ be a positive real number smaller than this minimum, and suppose that $\mathbf{F}$ is any regular δ-family of blocks whose union contains A. Let $\mathbf{F}'$ be the subfamily of $\mathbf{F}$ consisting of the blocks contained in A°. By the choice of δ, every member of $\mathbf{F}$ meeting a member of $\mathbf{G}_\sigma$ belongs to $\mathbf{F}'$. Consequently, the union of $\mathbf{G}_\sigma$ is contained in the union of $\mathbf{F}'$. Therefore, the volume of the union of $\mathbf{F}'$ is no smaller than $\sigma^n(V(A) - \varepsilon)$.

It will be convenient to use the following notation. Let $\mathbf{F}$ be an arbitrary family of subsets of $\mathbf{R}^n$. Then $IN_A(\mathbf{F})$ denotes the subfamily of $\mathbf{F}$ consisting of the members that are contained in A°, while $ON_A(\mathbf{F})$ denotes the family of all members of $\mathbf{F}$ that have a point in common with A. Combining our last result with the previous one concerning outer measure, we obtain the following conclusion.

Suppose that A is a closed bounded subset of $\mathbf{R}^n$ having volume. For every positive real number ε, there is a positive real number τ such that, if $\mathbf{F}$ is any *regular τ-family of blocks whose union contains A, then both the volume of the union of $IN_A(\mathbf{F})$ and the volume of the union of $ON_A(\mathbf{F})$ differ from the volume of A by no more than ε.*

5. Let C be a closed, bounded and convex subset of $\mathbf{R}^n$. We assume that $n > 1$, and we regard $\mathbf{R}^n$ as the Cartesian product $\mathbf{R} \times \mathbf{R}^{n-1}$. For every real number ρ, we let $C(\rho)$ denote the set of points p of $\mathbf{R}^{n-1}$ such that (ρ, p) belongs to C. Since C is convex, the set of real numbers ρ for which $C(\rho)$ is non-empty is a (possibly degenerate) interval $[\alpha, \beta]$. Evidently, each set $C(\rho)$ is closed, bounded and convex, so that it has $(n-1)$-volume. We define the non-negative real valued function f on $[\alpha, \beta]$ by setting $f(\rho)$ equal to the $(n-1)$-volume of $C(\rho)$. We call f the *cross-section function* of C, and we shall prove that *the cross-section function is continuous.*

Let $\mathbf{K}$ denote the family of all cubical blocks of edge length 1 whose vertices are of the form $(z_1, \ldots, z_n)$, where each z_i is an integer. If σ is a positive real number and p is a point of $\mathbf{R}^n$, then the cubical blocks of the form $p + \sigma K$, with K in $\mathbf{K}$, are of diameter $\sigma\sqrt{n}$ and have pairwise disjoint interiors. We refer to the family of these blocks as a *σ-grid* of $\mathbf{R}^n$.

Given a positive real number ε and a cross-section $C(\rho)$, choose σ such that the pair $(\sigma\sqrt{n-1}, C(\rho))$ has the property expressed for (τ, A) at the end of Section 4. Let $\mathbf{G}$ be a σ-grid of $\mathbf{R}^{n-1}$. By an argument used before (cf. the last proof in Section 4), we see that there is a positive real number δ such that, if G is a member of $\mathbf{G}$ and $G \cap C(\rho) = \varnothing$, then the distance between (ρ, G) and C is greater than δ.

Now let ρ' be a point of $[\alpha, \beta]$ such that $|\rho - \rho'| \leqq \delta$. Let G be a member of $\mathbf{G}$ meeting $C(\rho')$. Then the distance between (ρ, G) and C is no greater than δ. Consequently, G meets $C(\rho)$. Thus, we have $ON_{C(\rho')}(\mathbf{G}) \subset ON_{C(\rho)}(\mathbf{G})$. By our choice of σ, the volume of the union of $ON_{C(\rho)}(\mathbf{G})$ is no greater than $f(\rho) + \varepsilon$. Since the volume of the union of $ON_{C(\rho')}(\mathbf{G})$ is at least $f(\rho')$, it follows from the above inclusion relation that $f(\rho') \leqq f(\rho) + \varepsilon$.

Now let us first deal with the case where $\alpha < \rho < \beta$. Let p be a point of the interior of $C(\rho)$. Using the assumption on ρ, we see that there are points u and v of C° such that $u = (\mu, u')$ and $v = (v, v')$, with u' and v' in $\mathbf{R}^{n-1}$ and $\mu > \rho > v$. For every point q in a neighborhood of p in $C(\rho)$, consider the line segments $[u,(\rho, q)]$ and $[v,(\rho, q)]$. We know from Section 3 that all the points of these line segments, with the possible exception of (ρ, q), belong to C°. Using this, together with the fact that C° is convex, we find that (ρ, p) must belong to C° also. Our conclusion is that $(\rho, C(\rho)^\circ) \subset C^\circ$. By a familiar argument, it follows now that there is a positive real number η such that, if G is member of $\mathbf{G}$ that is contained in the interior of $C(\rho)$, then the distance between (ρ, G) and the boundary of C is greater than η. Let ρ' be a point of $[\alpha, \beta]$ such that $|\rho - \rho'| \leqq \eta$, and let G be a member of $\mathbf{G}$ that lies in the interior of $C(\rho)$. Since the distance between (ρ', G) and (ρ, G) is no greater than η, it is clear from our choice of η that (ρ', G) lies in the interior of C, which implies that G lies in the interior of $C(\rho')$. Our conclusion is that $IN_{C(\rho)}(\mathbf{G}) \subset IN_{C(\rho')}(\mathbf{G})$.

In particular, this shows that $f(\rho') \geqq f(\rho) - \varepsilon$. From this and the volume inequality obtained before, it is clear that the cross-section function is continuous on the interior of the interval $[\alpha, \beta]$.

Actually, an inequality like this last one can be established by the following simple argument, which does not depend on any restriction of ρ. Let ρ_1 be any point of $[\alpha, \beta]$, other than ρ. Fix a point p in $C(\rho_1)$. For every real number ρ' in $[\rho, \rho_1]$, the set

$$(\rho', (\rho - \rho_1)^{-1}((\rho - \rho')p + (\rho' - \rho_1)C(\rho)))$$

is contained in $(\rho', C(\rho'))$, because of the convexity of C. It follows from this that

$$f(\rho') \geqq \left(\frac{\rho' - \rho_1}{\rho - \rho_1}\right)^{n-1} f(\rho).$$

Therefore, $f(\rho') \geqq f(\rho) - \varepsilon$ as soon as ρ' is close enough to ρ. This completes the proof of the continuity of the cross-section function on $[\alpha, \beta]$.

Let f denote the cross-section function of C. Since f is continuous, the integral $I_{[\alpha,\beta]}(f)$ is defined. Now we are in a position to show that *the volume of C is equal to the integral $I_{[\alpha,\beta]}(f)$ of the cross-section function.*

Assuming, as we may, that $\alpha < \beta$, let α_1 and β_1 be real numbers such that $\alpha < \alpha_1 < \beta_1 < \beta$. Let ε be a given positive real number. From the inequalities we have proved above for interior points ρ of $[\alpha, \beta]$, we know that, for every ρ in $[\alpha_1, \beta_1]$, there is a positive real number $s(\rho)$ satisfying the following condition. If $\mathbf{G}$ is a σ-grid in $\mathbf{R}^{n-1}$, with $\sigma \leqq s(\rho)$, and ρ' is a point of $[\alpha, \beta]$ such that $|\rho' - \rho| \leqq s(\rho)$, then the volume of the union of $ON_{C(\rho')}(\mathbf{G})$ differs from $f(\rho)$ by no more than ε. We claim that there is a positive real number η that satisfies the condition on $s(\rho)$ for every ρ in $[\alpha_1, \beta_1]$. In fact, if such an η does not exist, the usual iterated bisection argument shows that there is a point ρ in $[\alpha_1, \beta_1]$ such that, for every positive integer t and every neighborhood W of ρ in $[\alpha_1, \beta_1]$, there is a point ρ^* in W such that $1/t$ cannot serve as $s(\rho^*)$. It is easy to see that this contradicts the existence of an $s(\rho)$ satisfying the above condition with $\varepsilon/2$ in the place of ε, so that our claim is established.

Now let $\mathbf{F}$ be a σ-grid of $\mathbf{R}^n$, with $\sigma \leqq \eta$. Then $\mathbf{F}$ determines a σ-grid $\mathbf{G}$ of $\mathbf{R}^{n-1}$ such that a set (γ, E), with γ in $\mathbf{R}$ and E contained in $\mathbf{R}^{n-1}$, is the intersection of a member of $\mathbf{F}$ with $(\gamma, \mathbf{R}^{n-1})$ if and only if E is a member of $\mathbf{G}$. Now, if one takes α_1 and β_1 to be sufficiently close to α and β, respectively, and if one counts the members of $ON_C(\mathbf{F})$ according to a grouping into layers between σ-spaced hyperplanes $(\gamma, \mathbf{R}^{n-1})$, one finds that the volume of the union of $ON_C(\mathbf{F})$ differs from $I_{[\alpha,\beta]}(f)$ by no more than

$$(\beta - \alpha)\varepsilon + 2\max(f)\sigma$$

The equality $V(C) = I_{[\alpha,\beta]}(f)$ follows readily from this result.

6. We need some notation to facilitate the description of our setting. Let B be a real interval of length $2\beta > 0$, and let β_0 denote the center of B. Let δ be a positive real number, and suppose we are given a continuous real-valued function f on $[0, \delta] \times B$. We let t stand for the identity map on $[0, \delta]$, or on subintervals $[0, \sigma]$, with $\sigma \leqq \delta$. If g is a continuous map from such a subinterval to B, then the composite function $f(t, g)$ is defined and continuous on $[0, \sigma]$, and the integral $I_{[0,\rho]}(f(t, g))$ is defined for every ρ in $[0, \sigma]$. We wish to treat the problem of finding such functions g with the property that, for every ρ in $[0, \sigma]$,

$$g(\rho) - \beta_0 = I_{[0,\rho]}(f(t, g))$$

The reader will recognize this as the fundamental problem concerning ordinary differential equations when he observes that the expression on the right is customarily expressed as $\int_0^\rho f(t, g(t))\,dt$, and that the above conditon is equivalent to the requirements that g be differentiable, that $g'(\rho) = f(\rho, g(\rho))$ and that $g(0) = \beta_0$.

The problem becomes tractable when the given function f satisfies the following (Lipschitz) condition. There is a (positive) real number γ such that, for every ρ in $[0, \delta]$,

$$|f(\rho, \beta_1) - f(\rho, \beta_2)| \leqq \gamma |\beta_1 - \beta_2|$$

whenever β_1 and β_2 lie in B.

Let μ be the maximum absolute value of f, and fix a positive real number η in $[0, \delta]$ such that $\mu\eta \leqq \beta$ and $\gamma\eta < 1$. Now let G denote the set of all continuous maps g from $[0, \eta]$ to B such that $g(0) = \beta_0$. We equip G with the norm $\| \ \|$, where $\|g\|$ is the maximum absolute value of g (cf. Section VI.4). For every g in G, define the function $S(g)$ on $[0, \eta]$ by the formula

$$S(g)(\rho) = \beta_0 + I_{[0,\rho]}(f(t, g))$$

Clearly, the absolute value of the integral on the right is at most $\mu\eta \leqq \beta$, whence it is clear that $S(g)$ belongs to G. If g_1 and g_2 belong to G, and ρ is a point in $[0, \eta]$, we have

$$\begin{aligned}|S(g_1)(\rho) - S(g_2)(\rho)| &\leqq I_{[0,\rho]}(|f(t, g_1) - f(t, g_2)|)\\ &\leqq I_{[0,\rho]}(\gamma|g_1 - g_2|) \leqq \gamma\eta\|g_1 - g_2\|\end{aligned}$$

which shows that

$$\|S(g_1) - S(g_2)\| \leqq \gamma\eta\|g_1 - g_2\|$$

Since $\gamma\eta < 1$, we can conclude from this result that, for every g in G, the iterated transforms $S^n(g)$ constitute a Cauchy sequence of functions (cf. Section VII.4). It is easy to see that this Cauchy sequence approaches a limit in G. Moreover, our inequality concerning S shows that this limit is the same for every g in G. The conclusion is that there is one and only one function g in G such that $S(g) = g$, i.e., such that g satisfies the requirements of our main problem.

It is of importance for numerical computation to have an estimate of the deviation of the solutions of approximations to a given differential equation from the correct solution. The following is an estimate of this kind. Suppose that h is a continuous map from $[0, \eta]$ to B such that, for every ρ in $[0, \eta]$,

$$|h(\rho) - \beta_0 - I_{[0,\rho]}(f(t, h))| \leqq \varepsilon\rho$$

where ε is some fixed positive real number. Then, if g denotes the above correct solution, one has

$$|g(\rho) - h(\rho)| \leqq (\varepsilon/\gamma)(\exp(\gamma\rho) - 1)$$

for every ρ in $[0, \eta]$.

In order to prove this, write

$$g(\rho) - h(\rho) = I_{[0,\rho]}(f(t, h)) + \beta_0 - h(\rho) + I_{[0,\rho]}(f(t, g) - f(t, h))$$

Using the assumptions on h and f, we see from this that

$$|g(\rho) - h(\rho)| \leqq \varepsilon\rho + \gamma I_{[0,\rho]}(|g - h|) \leqq \varepsilon\rho + \gamma\|g - h\|\rho.$$

Now suppose that it has already been shown, for some positive index n, that (as the above says for $\eta = 1$)

$$|g(\rho) - h(\rho)| \leqq (\varepsilon/\gamma)[\gamma\rho + \cdots + (\gamma\rho)^n/n!] + (\gamma\rho)^n\|g - h\|/n!$$

Substituting this in the integral above, we find

$$|g(\rho) - h(\rho)| \leqq \varepsilon\rho + \gamma I_{[0,\rho]}((\varepsilon/\gamma)[\gamma\rho + \cdots + (\gamma\rho)^n/n!] + (\gamma\rho)^n\|g - h\|/n!)$$

Carrying out the integration on the right yields exactly the preceding inequality with $n + 1$ in the place of n. Thus, the inequality holds for all positive indices n. Making n large, we obtain the result we set out to prove.

Exercises

1. Let us agree that a Z-point of $\mathbf{R}^n$ is a point of the form $(\tau_1, \ldots, \tau_n)$ where each τ_i is an integer. Suppose that C is a closed, bounded and convex subset of $\mathbf{R}^n$ such that $-C = C$ (i.e., C is symmetric with respect to 0) and the volume of C is greater than 2^n. Show that C contains at least one Z-point other than 0. By considering $(1 + \rho)C$ for small positive real numbers ρ, extend the result to the case where the volume of C is equal to 2^n. [Let $\mathbf{K}$ be a 1-grid of $\mathbf{R}^n$, in the sense of Section 5. Write $2^{-1}C$ as the union of the family of its intersection with the members K of $\mathbf{K}$. Fix a member K_0 of $\mathbf{K}$, and note that every member K is of the form $t + K_0$, where t is a Z-point, and that the volume of $2^{-1}C \cap K$ equals that of $(-t + 2^{-1}C) \cap K_0$. Show that the initial assumption on the volume of C implies that the intersections $2^{-1}C \cap K$ cannot be pairwise disjoint, and deduce the result from this fact.] Finally, show that the bound 2^n for the volume is best possible.

2. For every positive integer n, let v_n denote the volume of the set of all points p in $\mathbf{R}^n$ such that $|p| \leqq 1$. Show that $v_{n+1} = 2v_n I_{[0,1]}((1 - t^2)^{n/2})$.

3. In Section 6, we took for granted that $I_{[0,\rho]}(t^q) = \rho^{q+1}/(q + 1)$ for every non-negative real number ρ and every non-negative integer q. Prove this result directly with the methods of this chapter. [There is no difficulty after having proved the result in the case where $\rho = 1$. In that case, write $I_{[0,1]}(t^q)$ as the limit of the sequence whose nth term is obtained from a partitioning of $[0, 1]$ into n subintervals of length $1/n$. The limit can be obtained with the use of the formula

$$\sum_{k=1}^{n} k(k + 1) \cdots (k + q - 1) = (q + 1)^{-1} n(n + 1) \cdots (n + q)$$

which is easily established by induction on n.]

4. Let α and β be real numbers, to be kept fixed. Prove that there is one and only one continuous real-valued function g on $\mathbf{R}$ such that $g(\rho) = \alpha + \beta\, I_{[0,\rho]}(g)$ for every real number ρ. Determine g directly from the construction in the existence proof of Section 6.

5. The aim of this exercise is to prove the following result. Suppose that S is a bounded subset of $\mathbf{R}^n$ that has volume. Let T be a linear endomorphism of $\mathbf{R}^n$. Then the image set $T(S)$ has volume, and the volume of $T(S)$ is the product of the volume of S and the absolute value of the determinant of T.

This follows quite easily from the result of Exercise V.3. An independent proof proceeds along the following lines.

If T is not invertible then $T(S)$ is contained in a proper sub vector space of $\mathbf{R}^n$, whence its n-volume is 0. On the other hand, the determinant of T is equal to 0, so that the result holds trivially in this case.

Thus, it suffices to deal with the case where T is invertible. In that case, show by induction on n that T is a composite of a finite family of linear automorphisms, each of which is of one of the following three types. Let $(e_1, \ldots, e_n)$ be the canonical basis of $\mathbf{R}^n$. An automorphism of type 1 sends e_1 to γe_1, where γ is a non-zero real number, and leaves the remaining basis elements fixed. An automorphism of type 2 interchanges two of the e_i's and leaves the remaining $n - 2$ basis elements fixed. An automorphism of type 3 leaves e_i fixed for each $i > 1$, and sends e_1 to $e_1 + e_2$.

It is easy to see that the result holds if T is of type 1 or type 2. In order to show that the result holds also if T is of type 3, note that it suffices to deal with the case where S is closed and convex, or even just with the case where S is a block. In that case, obtain the required result by applying Section 5.

Project

In the notation of the beginning of Section 6, explore the following computational procedure for obtaining an approximation g to a solution of the problem. Choose a large positive integer n, define $g(0) = \beta_0$ and, recursively for $k = 0, 1, \ldots, n$,

$$g((k+1)n^{-1}\delta) = g(kn^{-1}\delta) + n^{-1}\delta f(kn^{-1}\delta, g(kn^{-1}\delta))$$

After extending this definition of g by linear interpolation, compare g with the correct solution by the method of Section 6. In particular, develop estimates yielding lower bounds for n ensuring that the error is acceptable. For a variety of functions f for which the solution can be obtained explicitly, run computer programs carrying out the above recursion and observe the dependence of the error on n.

CHAPTER IX

The Sphere in 3-Space

1. Let V be a Euclidean space, W an m-dimensional sub $\mathbf{R}$-space of V, p a point of V. Then the subset $p + W$ is called an m-dimensional *affine subspace* of V. Suppose that f is an injective linear map from $\mathbf{R}^m$ to W, and let B be a block in $\mathbf{R}^m$. Then $f(B)$ is a parallelepiped in W, and therefore has m-volume. As is evidently appropriate, we define the m-volume of the subset $p + f(B)$ of the affine space $p + W$ to be the m-volume of $f(B)$. We refer to $p + f(B)$ as an *affine m-patch* in V.

One extends the definition of m-volume to more general sets by approximating them, generally in rather subtle ways, with unions of finite families of affine m-patches. In such approximations, the affine subspace containing a patch must be allowed to vary with the patch. We have already met the simplest example of such an approximation. Namely, a rectifiable image of a simple continuous path in $\mathbf{R}^2$, such as a circular arc. In this case, $m = 1$, and the affine 1-patches are the line segments joining successive points of a progression of points on the path.

A general setting in which such approximations succeed is as follows. Let D be an open subset of $\mathbf{R}^m$, and let f be a map from D to a Euclidean space V. One says that f is *differentiable* at a point p of D if there is a linear map f'_p from $\mathbf{R}^m$ to V that approximates f in a neighborhood of p in the following sense. For every point q of D, one has

$$f(q) = f(p) + f'_p(q - p) + |q - p| e_p(q)$$

where $e_p(q)$ is a point of V approaching the origin as q approaches p. The linear map f'_p is the (geometric) *derivative* of f at p. Let us assume that f is differentiable at every point of D, that each f'_p is injective and that f'_p depends continuously on p, in the sense that, for every point t of $\mathbf{R}^n$, the map from D

to V sending each p to $f'_p(t)$ is continuous. Finally, let B be a block of $\mathbf{R}^m$ that is contained in D, and suppose that f is injective from B to V.

Now consider regular τ-families of blocks whose unions coincide with B (cf. Section VIII.2). If $\mathbf{F}$ is such a family and F is a member of $\mathbf{F}$, let p denote the center of F, and let $\pi(F)$ denote the affine m-patch $f(p) + f'_p(-p + F)$ in V. When τ is small, the union of the family of patches $\pi(F)$ with F ranging over $\mathbf{F}$ approximates $f(B)$, and the sum of the m-volumes of these patches approaches a certain non-negative real number depending only on $f(B)$ as $\mathbf{F}$ ranges over a sequence in which τ approaches 0. This limit serves as the m-volume of the set $f(B)$.

In this way, the m-volume appears as an integral, as follows. For each point p of D, let $f^*(p)$ denote the m-volume of $f'_p(E_m)$ in V, where E_m is the unit block in $\mathbf{R}^m$ whose vertices are the 2^m sums formed from the canonical basis $(e_1, \ldots, e_m)$ of $\mathbf{R}^m$. The function f^* so defined may be called the *numerical derivative* of f. The value $f^*(p)$ can be computed from the m-tuple $(f'_p(e_1), \ldots, f'_p(e_m))$ by the formula established in Exercise V.3. Consequently, our assumptions on f imply that f^* is continuous. If p is the center of a block F as above, then the m-volume of the affine patch $f(p) + f'_p(-p + F)$ is equal to the product of $f^*(p)$ and the volume of F. Therefore, *the m-volume of $f(B)$ is equal to the integral $I_B(f^*)$.* Usually, this is taken as the definition of m-volume. It can be shown that, actually, this integral depends only on $f(B)$, not on the particular choice of f. Finally, we remark that, if g_p is a distance preserving linear map from $f'_p(\mathbf{R}^m)$ to $\mathbf{R}^m$, then $f^*(p)$ equals the absolute value of the determinant of $g_p \circ f'_p$.

As an example, consider a right circular cylinder of radius ρ and height η in $\mathbf{R}^3$. For notational convenience, let us view $\mathbf{R}^3$ as the Cartesian product $\mathbf{C} \times \mathbf{R}$. The points of our cylinder may then be written $(\rho \exp(\sigma i), \tau)$, with σ in $[0, 2\pi]$ and τ in $[0, \eta]$. We calculate one half of the surface area (2-volume) of our cylinder by using the map f from $\mathbf{R}^2$ to $\mathbf{R}^3$ given by $f(\sigma, \tau) = (\rho \exp(\sigma i), \tau)$. Writing $\sigma + \alpha$ for σ and $\tau + \beta$ for τ, and using the power series presentation of exp, we see that f is differentiable, and that $f'_{(\sigma,\tau)}(\alpha, \beta) = (\rho \exp(\sigma i)\alpha i, \beta)$. One half of our cylinder is the image by f of the block $[0, \pi] \times [0, \eta]$. A simple direct computation shows that f^* is the constant function with value ρ. Consequently, the surface area of the half cylinder is $I_{[0,\pi]\times[0,\eta]}(\rho) = \pi\eta\rho$.

Observe that the affine patches obtained from our map f are tangential to the cylinder. On the other hand, the straightforward generalization of the approximation to curves by chains of chords would be the approximation to surfaces by assemblies of triangles whose vertices lie on the surface.

It is interesting to note that this approximation procedure is inappropriate (unless it is subjected to somewhat unnatural constraints). In our above description of the cylinder, consider the triangles with vertices

$$f(\pi(k-1)/n, \tau),\ f(\pi k/n, \tau + \beta),\ f(\pi(k+1)/n, \tau)$$

or

$$f(\pi k/n, \tau + \beta),\ f(\pi(k+1)/n, \tau),\ f(\pi(k+2)/n, \tau + \beta)$$

where τ and β are fixed positive real numbers, n is a fixed positive integer, and k ranges over the integers from 1 to $2n - 1$. These triangles fit together to form a faceted ribbon riveted at the above vertices to the part of the cylinder between the "heights" τ and $\tau + \beta$. As β approaches 0, while n is kept fixed, the direction of the normal to each triangle approaches that of the axis of the cylinder. It follows that the sum of the areas of the family of all triangles used for approximating the cylinder (taking $\beta = \eta/m$, with m a large positive integer, and letting τ ranges over the integer multiples $q\beta$, with $q = 0, \ldots, m - 1$) increases without bound as β approaches 0. This shows that there are "approximations" to the cylinder by assemblies of triangles of arbitrarily small diameter whose total areas are (simultaneously) arbitrarily large.

2. Now let us consider the sphere $S_\rho(0)$ in $\mathbf{R}^3$ whose center is the origin 0 and whose radius is ρ. This is the set of all points p in $\mathbf{R}^3$ such that $|p| = \rho$. Again, let us regard $\mathbf{R}^3$ as the Cartesian product $\mathbf{C} \times \mathbf{R}$. Then the points of $S_\rho(0)$ may be written as pairs $(\rho\cos(\gamma)\exp(\sigma i), \rho\sin(\gamma))$, with γ in $[-\pi/2, \pi/2]$ and σ in $[0, 2\pi]$. Fix σ arbitrarily, fix γ so that $|\gamma| < \pi/2$, and let μ be a positive real number such that $\gamma + \mu$ and $\gamma - \mu$ belong to $[-\pi/2, \pi/2]$. Let P and Q be the rays whose source is the origin of $\mathbf{R}^3$ and which contain the points of $S_\rho(0)$ obtained by replacing γ with $\gamma - \mu$ and $\gamma + \mu$ in the above, respectively. One sees easily from a drawing that if p belongs to P and q belongs to Q and $|p| = |q| = \varepsilon$, say, then

$$|p - q|\cos(\gamma) = \varepsilon(\sin(\gamma + \mu) - \sin(\gamma - \mu))$$

Slicing $S_\rho(0)$ into ribbons by planes corresponding to closely spaced values of γ, one sees from this that $S_\rho(0)$ has 2-volume, and that this is equal to that of the cylinder of radius ρ and height 2ρ, i.e., that the surface area of $S_\rho(0)$ is equal to $4\pi\rho^2$.

Let T be an orthogonal linear automorphism of $\mathbf{R}^3$. For general reasons, and essentially by the result of Exercise V.2, T preserves 2-volume (as well as 1-volume and 3-volume), i.e., if K is a subset of $\mathbf{R}^3$ having 2-volume, then $T(K)$ has 2-volume and its 2-volume equals that of K. Evidently, $T(S_\rho(0)) = S_\rho(0)$. This remark is of importance for the geometry of the sphere. In particular, we shall use it in establishing the basic relation between angles and areas on the sphere.

Recall that a *great circle* on a sphere is the intersection of the sphere with a plane containing the center. The intersection of every pair of distinct great circles is a pair of antipodal points. If p is any point of our sphere, we denote its antipode by p^*. Of course, if our sphere is $S_\rho(0)$, then $p^* = -p$. Now let P and Q be two distinct planes containing the center of the sphere. Denote the two antipodal points of the sphere that lie on the line $P \cap Q$ by x and

x^*. If P and Q are removed from the 3-dimensional Euclidean space containing our sphere, there remain four pairwise disjoint regions whose intersections with the sphere are four *lunes*, each of which is bounded on the sphere by a pair of semi great circles joining x and x^*. There are several equivalent choices of a numerical measure for the *angle at* x that is represented by such a lune. One is as follows. Let R be the plane through the center of the sphere and orthogonal to $P \cap Q$. The intersection of R with our lune is a circular arc. If the length of this arc is α and the length of the great circle is η (in the case of $S_\rho(0)$, we have $\eta = 2\pi\rho$) then the measure of the angle may be taken to be $2\pi\alpha/\eta$.

On the other hand, it is easy to see that every lune has 2-volume. Using the additivity of 2-volume and the fact that rotations of the 3-space around $P \cap Q$ as axis are orthogonal linear automorphisms, we see that, if σ is the surface area of our sphere (in the case of $S_\rho(0)$, we have $\sigma = 4\pi\rho^2$) and μ is the area of our lune, then $\mu/\sigma = \alpha/\eta$, so that we may also use $2\pi\mu/\sigma$ as a numerical measure of the angle represented by the lune.

Now let us consider a triangle on our sphere, with vertices a, b, c and arcs of great circles as edges. When this triangle is deleted from the sphere, there remain two disjoint regions, such that two points p and q belong to the same region if and only if they can be joined by a finite chain of great circular arcs not meeting the triangle. We assume that our triangle is non-degenerate, in the sense that a, b and c do not all lie on one great circle. Then exactly one of our two regions is contained in a hemisphere. First, we deal with the case where this region is regarded as the "interior" of the triangle.

In this case, we begin under the additional assumption that each edge is contained in a *semi* great circle. Then, the interior, J say, is the intersection of three lunes A, B and C, where the boundary of A consists of the semi great circle from a via b to a^* and the semi great circle from a via c to a^*, and the boundaries of B and C are obtained mutatis mutandis. We claim that the surface area of $A \cup B \cup C$ is equal to $\sigma/2$. In order to see this, let E denote the complement of J in A. Let E^* be the set of all antipodes e^* of points e of E. Since the map sending each point p of the containing 3-space to $-p$ is an orthogonal linear automorphism, we know that E^* has 2-volume, and that this, the area of E^*, is equal to that of E. From an appropriate drawing, it is easy to see that the three sets $B \cup C$, E and E^* are pairwise disjoint, that $A \cup B \cup C = E \cup B \cup C$ and that $E^* \cup B \cup C$ is precisely a hemisphere from which the bounding great circle and two additional parts of great circles have been deleted. Consequently, the surface area of $A \cup B \cup C$ is equal to $\sigma/2$.

Now let α, β, γ denote the numerical measures of the angles of our triangle, and let μ stand for the area of the interior J. Then the areas of A, B and C are equal to $\alpha\sigma/(2\pi)$, $\beta\sigma/(2\pi)$ and $\gamma\sigma/(2\pi)$. The sum of these areas is equal to the area of $A \cup B \cup C$ plus twice the area of J. Therefore, our above result yields $(\alpha + \beta + \gamma)\sigma = 2\pi(2\mu + \sigma/2)$, or $\alpha + \beta + \gamma = 4\pi\mu/\sigma + \pi$.

Still under the assumption that the interior J is contained in a hemisphere, it remains to dispose of the case where one edge of the triangle, say the edge joining a to b, is not contained in a semi great circle. In this case, we may apply the above result to the triangle resulting from replacing this edge with the other part of the same great circle joining a to b. This triangle has angles $\pi - \alpha$, $\pi - \beta$, $2\pi - \gamma$, and its interior has area $\sigma/2 - \mu$. One sees directly that our above result for this changed situation yields the same result for the given case that we had under the additional assumption.

It remains to consider the case where the other region determined by the triangle is regarded as the interior. In this case, we apply the above to the complementary interior. The effect of the interchange of interiors is to change the angular measures to $2\pi - \alpha$, $2\pi - \beta$, $2\pi - \gamma$, while the area is changed to $\sigma - \mu$. Again, it appears upon substitution above that the same formula holds as in the case considered first. Our result may be expressed as follows.

Let S be a sphere of radius ρ, and let T be a triangle on S, with specified interior. If α, β and γ are the numerical measures of the angles of T, and $s(T)$ denotes the area of the interior of T, then $\alpha + \beta + \gamma = \rho^{-2}s(T) + \pi$.

3. Generalizing triangles on the sphere (with specified interior), we introduce (simple) *spherical p-gons* for every $p > 2$. Such a p-gon has p vertices and p edges, which are great circular arcs. Its complement on the sphere consists of two disjoint regions, one of which is specified as the *interior* of the p-gon. A spherical p-gon is called *regular* if all its vertices lie on one plane, all its edges are of the same length and all its angles have the same numerical measure (the first condition is actually a consequence of the other two). We wish to determine the presentations of the sphere as unions of families of pairwise congruent regular p-gons whose interiors are pairwise disjoint. Let us call such a presentation a *regular partition* of the sphere.

Consider such a presentation in which the total number of p-gons is n, and where the number of p-gons sharing a vertex is k. Being familiar with oranges, we assume that $p > 2$. Consider one of the p-gons. From each of its vertices draw an arc of a great circle to its interior center. Now the p-gon with its interior appears as the union of p congruent spherical triangles with their interiors. If α is the measure of each angle of the p-gon, the measures of the angles of each of these spherical triangles are $\alpha/2$, $\alpha/2$ and $2\pi/p$. In order to simplify notation, we take the radius of our sphere to be equal to 1. Then the area of each of these triangles is equal to $4\pi/pn$. Now the result of Section 2 concerning the sum of the angles of a spherical triangle yields the equality

$$\alpha + 2\pi/p = \pi + 4\pi/pn$$

On the other, the sum of the angles at each vertex must be equal to 2π, so that we have $k\alpha = 2\pi$. Substituting this above and rearranging the result, we obtain the relation

$$n(2p - (p - 2)k) = 4k \qquad (*)$$

This shows that we must have $(p - 2)k < 2p$, which we may write as $k < 2/(1 - 2p^{-1})$. We see from this that if $p \geqq 6$ then k must be equal to 2, whence $n = 2$. The corresponding regular partitions present the sphere as the union of two hemispheres, which are viewed as p-gons by marking p equally spaced vertices on the common great circle. Actually, such a hemispherical partition exists also for $p = 2, 3, 4$ and 5, and we shall not include them again in the enumeration that follows. The remaining possibilities can be read off from (∗) as follows, where the regular polytope corresponding to a regular partition is the configuration in $\mathbf{R}^3$ resulting by replacing each spherical p-gon with the plane p-gon having the same vertices.

p	k	n	corresponding regular polytope
5	3	12	dodecahedron
4	3	6	cube
3	5	20	icosahedron
3	4	8	octahedron
3	3	4	tetrahedron

The relation (∗) is somewhat miraculous in three respects. First, the superficially permissible values of p and k all lead to *integer* values of n. Second, all the triples (p, k, n) allowed by (∗) actually correspond to regular partitions of the sphere (but it would take us too far afield to prove this here). Third, the following purely combinatorial consideration, involving no volume, area, length or angles, leads to the same relation (∗) as our above consideration of angles and areas.

Let us consider general partitions of the sphere by not necessarily regular p-gons, where we allow p to vary within a partition. Call the interiors of the p-gons the *faces* of the partition. An *edge* may now be understood to be an arc with one or two vertices as extremities, which are *not* regarded as points of the edge. Such an arc need not be assumed to be an arc of a great circle, but merely the image of a continuous map from [0, 1] to the sphere that is injective on the interior of [0, 1] and maps each of 0 and 1 to a vertex. Our assumptions are that the faces are pairwise disjoint, the edges are pairwise disjoint and that each edge belongs to the boundary of exactly two faces. The *Euler characteristic* of such a scheme of faces, edges and vertices is defined as the number of vertices minus the number of edges plus the number of faces. The critical fact is that, *for every scheme coming from a partition of the sphere, the Euler characteristic is equal to* 2.

This is seen as follows. For every pair of partitions, one may construct a common *refinement*, whose faces, edges and vertices are obtained from the superposition of the partitions in the evident way. Any refinement of a partition can be constructed in elementary steps of the following two types: (1) the introduction of a new vertex, cutting an edge into two new edges; (2) the introduction of a chain of edges and of their extremities as vertices (if new), where all the edges lie in a single face and where the chain cuts that face into two new faces. One sees immediately that neither (1) nor (2) changes the value of the Euler characteristic. Thus, the Euler characteristic is the same for all partitions of the sphere. Finally, the partition corresponding to the tetrahedron has four vertices, six edges and four faces, showing that the Euler characteristic is equal to two.

Now let p, k and n have the same meanings as before. Then the number of vertices is pn/k, the number of edges is $pn/2$ and the number of faces is n. Consequently, our result about the Euler characteristic yields the relation

$$pn/k - pn/2 + n = 2$$

which is merely another form of the relation $(*)$.

4. Let S denote the unit sphere $S_1(0)$ in $\mathbf{R}^3$. If we regard S as a metric space in its own right, with the distance function inherited from that of $\mathbf{R}^3$, we encounter the problem of the determination of the congruences of S, i.e., the bijective maps from S to S preserving Euclidean distance. Evidently, the map associating with each orthogonal linear automorphism of $\mathbf{R}^3$ its restriction to S is an injective group homomorphism from the group of orthogonal linear automorphisms of $\mathbf{R}^3$ to the group of congruences of S. We claim that this homomorphism is actually also surjective.

In order to see this, let T be a congruence of S. Every point of $\mathbf{R}^3$ other than the origin 0 can be written in one and only one way in the form αa, where α is a positive real number and a is a point on S. We extend the definition of T to $\mathbf{R}^3$ by putting $T(\alpha a) = \alpha T(a)$ and $T(0) = 0$. Evidently, this extended map T is a bijective map from $\mathbf{R}^3$ to $\mathbf{R}^3$ leaving the origin fixed. Also it is clear from the definition that $|T(p)| = |p|$ for every point p. Now let p and q be distinct points other than 0. Write $p = \alpha a$, $q = \beta b$, as above. Then we have

$$|T(p) - T(q)|^2 = \alpha^2 + \beta^2 - 2\alpha\beta T(a) \cdot T(b)$$

From the fact that $|T(a) - T(b)| = |a - b|$, we see that $T(a) \cdot T(b) = a \cdot b$. Substituting this above, we see that $|T(p) - T(q)| = |p - q|$. We know that every distance preserving bijection from $\mathbf{R}^3$ to $\mathbf{R}^3$ keeping 0 fixed is an orthogonal linear automorphism, so that our above claim is now established.

Our result is that *the restriction map is an isomorphism from the group of all orthogonal linear automorphisms of* $\mathbf{R}^3$ *to the group of all congruences of* S.

We wish to examine this group of congruences in more detail. Let H and K be hyperplanes (containing the origin) in an $\mathbf{R}^n$, with $n > 1$. We know that

$$\dim(H + K) + \dim(H \cap K) = \dim(H) + \dim(K)$$

(cf. Exercise IV.4). Let ρ and σ denote the reflections of $\mathbf{R}^n$ determined by H and K. Evidently, $\dim(H + K)$ is either $n - 1$ or n. In the first case, $H = K$, so that $\rho = \sigma$. In the second case, the above relation gives $\dim(H \cap K) = n - 2$. Let W be the orthogonal complement of $H \cap K$ in $\mathbf{R}^n$, as defined in Exercise IV.5. Then $\dim(W) = 2$. Since both ρ and σ leave the points of $H \cap K$ fixed, it is clear that they map W onto itself. From the fact that ρ and σ are reflections, it follows that their restrictions to W have determinant -1. Consequently, the determinant of the restriction of $\rho\sigma$ to W is equal to 1. We know that this implies that this restriction is a rotation (cf. Section IV.3). We may express our result as follows.

If $n > 1$, then the composite of every pair of reflections of $\mathbf{R}^n$ acts as a rotation on some 2-dimensional subspace and leaves the points of the orthogonal complement fixed.

Now let us apply this in the case where $n = 3$. We know that then every orthogonal linear automorphism is a composite of at most three reflections. With the above, this gives the conclusion that every orthogonal linear automorphism of $\mathbf{R}^3$ is either a plane rotation or a plane rotation followed by a reflection.

We can obtain a more precise result, as follows. Let T be an orthogonal linear automorphism of $\mathbf{R}^3$. Then the characteristic polynomial of T, as defined at the end of Chapter IV, is of degree 3, so that it has a real root. As was shown at the end of Chapter IV, this implies that there is a 1-dimensional subspace, A say, of $\mathbf{R}^3$ such that $T(A) = A$. Let P denote the orthogonal complement of A in $\mathbf{R}^3$. Then it follows from the fact that T is orthogonal that $T(P) = P$. Now the restriction of T to A is either the identity map or the negative of the identity map. In the first case, the determinant of the restriction of T to P is equal to the determinant of T. In the second case, these two determinants are the negatives of each other. We are interested in the case where the determinant of T is negative. Then, if T leaves the elements of A fixed, the determinant of its restriction to P is negative, so that this restriction is a reflection leaving the points of some 1-dimensional subspace, L say of P fixed. Evidently, T is therefore the reflection of $\mathbf{R}^3$ leaving the points of the plane $A + L$ fixed. If T maps each element a of A to $-a$, then the determinant of the restriction of T to P is positive, whence that restriction is a rotation of P. Finally, if the determinant of T is positive then T cannot be a composite of an odd number of reflections, so that T is a rotation of $\mathbf{R}^3$. Our conclusion is the following.

Let T be an orthogonal linear automorphism of $\mathbf{R}^3$. If the determinant of T is positive (*actually*, 1) *then T is a rotation around some line through the origin as axis. If the determinant of T is negative* (*actually*, -1) *then T is the composite of such a rotation with the reflection whose plane of fixed points is the orthogonal complement of the axis of that rotation.*

Returning to our sphere S, we are now in a position to describe all possible sets of fixed points for a given congruence of S. If the congruence is a non-trivial rotation, then the fixed point set is evidently a pair of antipodal points. If the congruence is a reflection, then the fixed point set is a great circle. If the congruence is the composite of a non-trivial rotation with a reflection whose plane of fixed points in $\mathbf{R}^3$ is the orthogonal complement of the axis of rotation, then it is easy to see that there are no fixed points.

5. The rotations of $\mathbf{R}^3$ constitute a normal subgroup of the group of all orthogonal linear automorphisms, and the corresponding factor group is of order 2. Indeed, the determinant function restricts to a surjective group homomorphism from the group of all orthogonal linear automorphisms of $\mathbf{R}^3$ to the multiplicative group consisting of 1 and -1, and the kernel of this group homomorphism is precisely the group of rotations. The computational control of the group of rotations of $\mathbf{R}^3$ is greatly facilitated by the use of *quaternions*, as follows.

The quaternions constitute a 4-dimensional $\mathbf{R}$-algebra containing $\mathbf{R}$. This $\mathbf{R}$-algebra is actually a (non-commutative) field, in the sense that the non-zero elements constitute a group with respect to the multiplication. The multiplication is defined by prescribing the products of the elements of a canonical basis $(1, i, j, k)$, where 1 is the identity element, as follows.

$$ij = k = -ji, \qquad jk = i = -kj, \qquad ki = j = -ik$$
$$i^2 = j^2 = k^2 = -1$$

Let $\mathbf{H}$ stand for this $\mathbf{R}$-algebra of quaternions. An $\mathbf{R}$-linear combination of i, j and k is called a *pure quaternion.* We denote the 3-dimensional $\mathbf{R}$-space of pure quaternions by V, so that, as an $\mathbf{R}$-space, $\mathbf{H} = \mathbf{R} + V$, with $\mathbf{R} \cap V = (0)$. It is convenient to introduce the linear automorphism of $\mathbf{H}$ that leaves the elements of $\mathbf{R}$ fixed and maps each pure quaternion to its negative. We indicate this linear automorphism by $*$, so that, if $u = \alpha + a$, where α is in $\mathbf{R}$ and a is in V, we have $u^* = \alpha - a$. One verifies immediately that $(uv)^* = v^*u^*$ for all quaternions u and v. Besides the quaternion multiplication, we shall use the inner product on $\mathbf{H}$ with respect to which $1, i, j, k$ are pairwise orthogonal unit vectors. For every quaternion u, we have $uu^* = u \cdot u$, as is seen directly from the definitions. Thus, if $u \neq 0$ then the reciprocal u^{-1} of u is given by $u^{-1} = (u \cdot u)^{-1}u^*$. Finally, one verifies directly that, *for all pure quaternions u and v, the sum $uv + u \cdot v$ is a pure quaternion.*

Now let u be any non-zero quaternion, and consider the $\mathbf{R}$-linear automorphism of $\mathbf{H}$ that sends each quaternion h to uhu^{-1}. We verify directly that $(uhu^{-1})^* = uh^*u^{-1}$. We know that a quaternion a is a pure quaternion if and only if $a^* = -a$. With the last statement, this shows that the linear automorphism of $\mathbf{H}$ corresponding to u restricts to a linear automorphism of V. Let us denote this linear automorphism of V by T_u. Note that $T_u = T_{\mu u}$ for every non-zero real number μ. Therefore, it suffices to consider T_u in the case where u is a *unit quaternion*, i.e., where $u \cdot u = 1$. Then we may write $u = \cos(\gamma) + \sin(\gamma)a$, where γ is a real number and a is a pure unit quaternion.

If s and t are pure quaternions, one has $st + ts = -2s \cdot t$, as is easily verified. Let us choose a pure unit quaternion b that is orthogonal to the above pure unit quaternion a. By virtue of what we have just stated, we have $ab + ba = 0$. Hence we find

$$(ab)^* = b^*a^* = (-b)(-a) = ba = -ab$$

which shows that ab is a pure quaternion. Moreover,

$$(ab)b + b(ab) = a(bb) - (bb)a$$

and the expression on the right is 0, because bb is real. It follows that $(ab) \cdot b = 0$. Similarly, $(ab) \cdot a = 0$. Finally, $(ab) \cdot (ab) = abb^*a^* = aa^* = 1$. Write c for ab. Then (a, b, c) is a basis of V consisting of pairwise orthogonal unit vectors. We have $ab = c = -ba$ and $ac = -b = -ca$. From this and the fact that $u^{-1} = \cos(\gamma) - \sin(\gamma)a$, we find that

$$T_u(b) = (\cos(\gamma)^2 - \sin(\gamma)^2)b + 2\sin(\gamma)\cos(\gamma)c$$

$$T_u(c) = -2\sin(\gamma)\cos(\gamma)b + (\cos(\gamma)^2 - \sin(\gamma)^2)c$$

while (evidently) $T_u(a) = a$. This shows that T_u is the rotation around $\mathbf{R}a$ as axis through an angle of numerical measure 2γ, the positive sense of rotation being from b toward c by the short route.

Now it is clear that *the map from the multiplicative group of unit quaternions to the group of rotations of* $\mathbf{R}^3$ *sending each* u *to* T_u *is a surjective group homomorphism whose kernel is the group consisting of* 1 *and* -1.

Exercises

1. Let f be the map from $[-\pi/2, \pi/2] \times [0, 2\pi]$ to $S_1(0)$ given by $f(\gamma, \sigma) = (\cos(\gamma)\cos(\sigma), \cos(\gamma)\sin(\sigma), \sin(\gamma))$ (cf. the beginning of Section 2). Show that $f^*(\gamma, \sigma) = cos(\gamma)$, and use this for determining the surface area of $S_1(0)$ as the integral of f^*.

2. Let p and q be distinct points on $S_1(0)$, and let C denote the semicircular arc containing p, q and p^*. For every point r of $S_1(0)$, let r° be the point of C that lies on the affine plane through r and perpendicular to $\mathbf{R}p$. Show that, for all points u and v of $S_1(0)$, one has $u \cdot v \leqq u^\circ \cdot v^\circ$, the equality holding only if u, v, p, p^* are coplanar. Deduce that if a rectifiable path on $S_1(0)$ whose extremities are p and q is of minimal length then it lies on a great circle. [For the first part, one may assume without loss of generality that $p = (0, 0, 1)$, and use the parametrization f of Exercise 1. In order to use the first result, note that $u \cdot v = \cos(\alpha)$, where α is the length of a great circular arc with extremities u and v.]

3. Prove that there are only three types of regular polygonal tiles with which the plane can be paved: squares, equilateral triangles or regular hexagons.

4. Regard the **R**-space **H** of quaternions as 4-dimensional Euclidean space, the inner product being as used in Section 5. Let U denote the group of unit quaternions. For a and b in U, define the linear automorphism $T_{a,b}$ of **H** by

$$T_{a,b}(h) = ahb^*$$

Show that the map from the direct product group $U \times U$ sending each (a, b) to $T_{a,b}$ is a surjective group homomorphism to the group of all orthogonal linear automorphisms of **H** with positive determinant (actually, 1), and that the kernel of this group homomorphism consists of the two elements $(1, 1)$ and $(-1, -1)$.

5. Let A be a finite-dimensional **R**-algebra that is a non-commutative field. Proceed as follows to show that A is isomorphic with **H**.

First, show that if c is any element of $A \backslash \mathbf{R}$ then the smallest sub **R**-algebra of A containing c is a field, F say, that is isomorphic with **C**, and every element of A that commutes with every element of F must belong to F.

Now we may suppose that $\mathbf{C} \subset A$. Let u be an element of $A \backslash \mathbf{C}$. Then $iu \neq ui$, and $iu - ui$ is a non-zero element, v say, of A such that $iv = -vi$. From this and the fact that v^2 belongs to $\mathbf{R} + \mathbf{R}v$, show that v^2 is equal to a negative real number. Next, show that the smallest sub **R**-algebra of A containing **C** and v may be identified with **H**, so that now we have $\mathbf{H} \subset A$.

Suppose that x and y are elements of A such that $x^2 = -1$, $y^2 = -1$ and the triple $(x, y, 1)$ is linearly independent over **R**. Show that $xy + yx$ belongs to $\mathbf{R} + \mathbf{R}x$ as well as to $\mathbf{R} + \mathbf{R}y$, whence it belongs to **R**. Use this result for showing that if a is an element of $A \backslash \mathbf{H}$ such that $a^2 = -1$ then one has $ia + ai = \rho$, $ja + aj = \sigma$ and $ija + aij = \tau$, where ρ, σ, τ belong to **R**. From this, deduce the contradiction $a + \frac{1}{2}(\rho i + \sigma j + \tau ij) = 0$.

Project

Design computer routines implementing quaternion arithmetic. There should be four functions of quaternions with quaternions as values: sum, negative, product, reciprocal. Note that such a facility contains complex number arithmetic. Accordingly, design an alternative quaternion arithmetic as a superstructure built on an available complex number facility.

Index